DESCRIPTIVE PAMPHLET

OF SOME OF

THE PRINCIPAL MINES AND PROSPECTS

OF

OURAY & SAN MIGUEL COUNTIES 1882-3

IN THE

SAN JUAN

GOLD AND SILVER REGION.

BY W. WESTON, MINER,

OF OURAY, COLO.

WESTERN REFLECTIONS PUBLISHING COMPANY®

Montrose, CO

ISBN-13: 978-1-932738-35-3
ISBN-10: 1-932738-35-5

First Edition
Printed in the United States of America

Cover and text design Laurie Goralka Design

Western Reflections Publishing Company®
219 Main Street
Montrose, CO 81401
www.westernreflectionspub.com

MAP OF OURAY & SAN MIGUEL MINING 1882-3

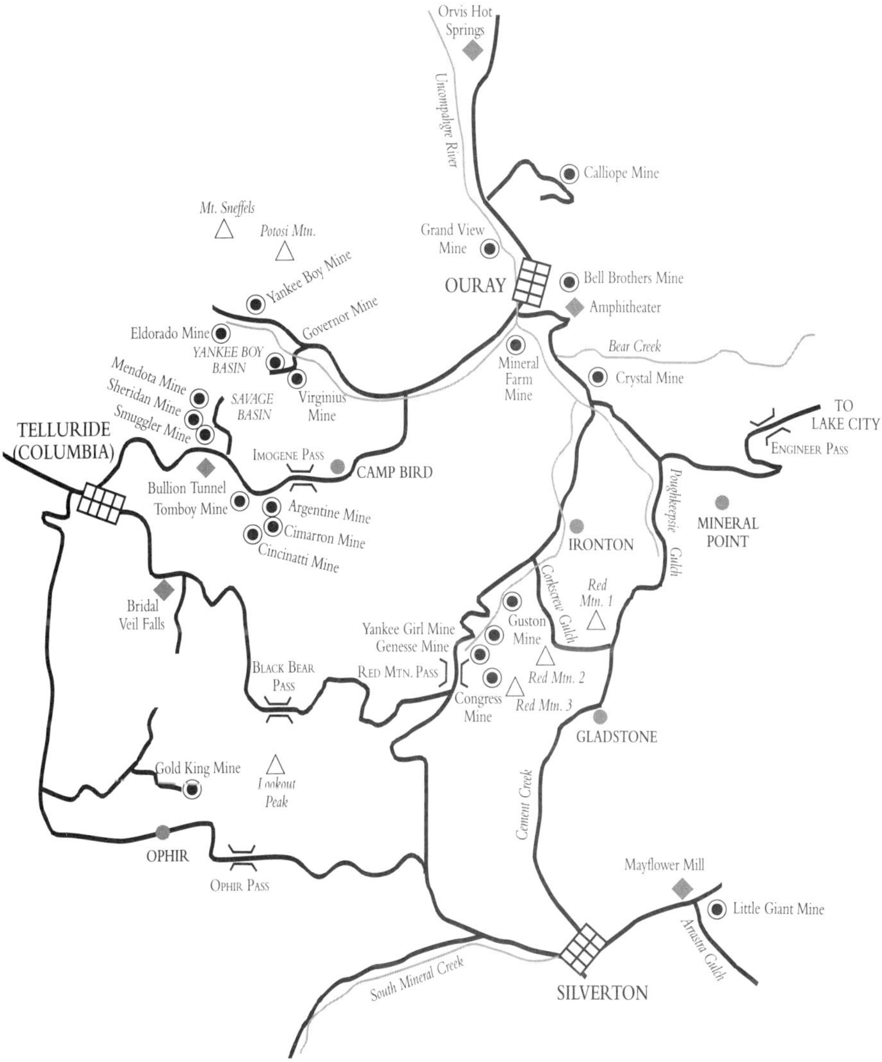

THAT PART OF THE SAN JUAN MINING AREA DESCRIBED BY WILLIAM WESTON.

FOREWARD

This exceedingly rare little book was published in 1883 by William Weston, who refers to himself (according to the practice of the day) as a "miner," but who was in fact by this time in his life a fully developed assayer, mining engineer, and promoter. Today the term "miner" would refer to one who worked daily for another in the mines, a practice which Weston did not keep.

William Weston was at one time the general European agent for the Kansas Pacific Railroad Company. When he succumbed to "gold and silver fever" in the early 1870s he did not rush to become a uneducated prospector (as so many men did) but rather went to England to attend the Royal School of Mines. He returned to the United States at the time of the "San Juan Excitement." Weston brought with him extensive assaying equipment and went to Silverton, Colorado, in 1876, where he was originally involved in road-building projects. By the spring of 1877 he was in Ouray County, working as a prospector with his partner George Barber in Yankee Boy and Imogene Basins, where they filed many important claims, including the Yankee Boy, Una, and Gertrude. For a little over three years the two men worked as "miners" developing their claims.

The San Juans had originally been prospected for gold in the 1860s — a decade late follow-up to the

WILLIAM WESTON'S OURAY HOME STILL STANDS AT 4TH STREET AND 5TH AVENUE. ALTHOUGH IT IS DIFFICULT TO TELL, WESTON MAY BE ONE OF THE TWO MEN STANDING BESIDE THE CARRIAGE.

Courtesy Western History Dept., Denver Public Library X-12793.

discovery of gold in California. Huge deposits of silver were overlooked in the hectic quest for gold. By 1876 large deposits of silver had been discovered, and after the rich silver strikes in Leadville, a mad scramble for silver was developing throughout Colorado — especially in the San Juans. Weston's Yankee Boy Mine produced good silver ore (copper, ruby silver, and brittle silver). In 1877 it reportedly produced $56,000. Weston and his partner leased the mine to F. B. Beaudry in 1878 for $58,000, which was a record price for a mining lease at the time. Beaudry shipped twenty tons of selected ore that year which assayed up to 1,700 ounces of silver per ton. In 1881 the mine was sold to a New York syndicate. Large amounts of sorted ore ran as high as 396 ounces of silver per ton and averaged 103 ounces of silver. The mine was rich enough to continue operating well into the twentieth century.

Other partners (the Reed brothers) eventually worked the silver veins of Weston's and Barber's Una and Gertrude Mines. Ironically the emphasis on gold had passed, and Weston, although a metallurgist, did not recognize the extremely valuable gold veins located directly outside the Gertrude and Una's silver vein. Weston and

THIS WELL-PREPARED (AND PROBABLY STAGED) PROSPECTOR IS HEADING INTO THE MOUNTAINS TO LOCATE VALUABLE MINERALS AND FILE A MINING CLAIM. PROSPECTORS DID ONLY MINIMAL DEVELOPMENT WORK, THEN SOLD THEIR CLAIMS AND MOVED ON TO LOOK FOR OTHER RICH DISCOVERIES.

Author's Collection.

Barber did extensive development on the Una and Gertrude, and Weston's cabin in Imogene basin became a local landmark, appearing on several maps of the time. Weston and Barber gave up partial ownership in the mines to gain further development, then they sold their remaining interest. Later the mines were sold to Thomas Walsh, who recognized their rich gold ore. Walsh used the claims as the basis for the fabulously rich Camp Bird Mine, which he eventually sold for six million dollars.

At the time this pamphlet was written, Ouray County encompassed the area that is now Ouray and San Miguel Counties, so although the title of the pamphlet was "A Descriptive Pamphlet of Some of the Principal Mines and Prospects of Ouray County, Colorado," it also contains details about many of the mines in what is now San Miguel County. The division into two counties was officially made on February 27, 1883, and is noted as an addendum at the very last page of the book. Most San Miguel mines developed slowly, and San Miguel ore continued to be shipped over the high mountain passes to Ouray, Silverton, or Lake City until the arrival of the Rio Grande Southern Railroad in Telluride in 1892.

Telluride was not the main settlement of San Miguel County at the time this pamphlet was written. San Miguel City had been laid out in 1876 about a mile west (down valley) of present-day Telluride. The town of Columbia, which eventually became Telluride, was started in 1878, but it was not until the early 1880s that its population began to approach that of San Miguel City. Both settlements had about 400 residents at the time of the writing of this pamphlet, as opposed to about a thousand residents in the more settled town of Ouray.

THE TOWN OF COLUMBIA WAS INCORPORATED ON SEPTEMBER 30, 1879, BUT THE NAME WAS CHANGED TO TELLURIDE IN 1887 BECAUSE A CALIFORNIA TOWN ALREADY USED THE NAME AND "CAL." AND "COL." WERE TOO DIFFICULT TO DISTINGUISH.

Author's Collection

The name of Columbia was not changed to Telluride until 1887.

Weston not only prospected but also used his marketing knowledge from the Kansas Pacific Railroad to help promote his properties and the San Juan mines in general. He regularly sent glowing reports (mostly justified) of the riches of the San Juan mines to the *Mining and Engineering Journal,* the *London Mining Journal*, and the *Denver Republican* newspaper.

Although many of his reports were self-motivated, Weston was largely responsible for the influx of money and prospectors into the San Juans in the late 1870s.

Much of this booklet was composed of material sent by Weston to newspapers in Denver in 1882. They were eventually compiled for the Denver Mining Exposition held in June of 1882. Denver held several mining expositions over the years to try to attract capital for Colorado mines. Weston's letters to newspapers were an attempt to attract capitalists (individuals or companies with money to invest). Mining in the San Juans was "hard rock" mining, which meant large expenditures of money were necessary to develop the mines. Transportation costs were also astronomical, which meant wagon roads and railroads were needed. All of this required large sums of money that the original prospectors didn't usually have. Weston continuously referred to the explosion of prospects that

San Miguel City was originally much bigger than Telluride, but by the time of this photograph (late 1880s) San Miguel City was rapidly declining while Telluride was booming as it was closer to the main mines in the area.

Courtesy Western History Dept., Denver Public Library X-13522.

Early reports of gold and silver in the San Juans sometimes grossly exaggerated how easy it was to "strike it rich." This *Harper's Weekly* drawing was said to illustrate how easy it was to spot silver veins in the mountains. It just wasn't so!

Author's Collection.

would occur if the local development and transportation problem could be solved.

Weston didn't just promote the local mines — he usually included rosy information on the towns of the area, and he often took the time to describe the area's great beauty and abundant wildlife. Weston's letters were heartedly received by the 1870s and 1880s newspapers, since they acted similarly to today's Chamber of Commerce. The local papers were expected to be strong advocates for the towns and communities that they served. It was through letters and newspaper exchanges that outsiders might learn of new discoveries in the mines, and the newspaper was an essential tool for promoting optimistic mining publicity for the would-be investor or speculator.

Weston was involved early in the Red Mountain Mining District, which included discoveries of incredible wealth. The first major discovery in the Red Mountains was not made until August of 1882, but at the time of Weston's writing the last part of this pamphlet (May 1883) it was evident that it would become such a major mining discovery that Weston chose to add information on the Red Mountain Mining District to his previous reports. Shortly after this little book was written he became the manager of Red Mountain's Guston Mine. Unfortunately there were allegations that Weston's mismanagement caused many problems for the English company that owned the mine. Weston disputed the claim but left as manager. A lawsuit resulted, and the mine was shut down until these matters were settled. The Guston eventually became one of the greatest silver mines in all of Colorado.

The east side of Ouray's Main Street was still pretty rough in 1880. Several of the commercial buildings are log cabins. But within a few years after the great Red Mountain mining discoveries, Ouray's future was secure, and mostly brick structures replaced the wooden ones.

Author's Collection.

Weston also became very involved with civic and business affairs in the City of Ouray, and in 1885 he was one of the incorporators of the Ouray Electric Light and Power Company. The use of alternating current had only recently been discovered and used in nearby Telluride, and it was so new that the Ouray electric company gave free electricity to a local saloon until the residents of Ouray became convinced of its usefulness. Weston also joined with a partner to venture into land speculation, creating a new subdivision on the Whipperwill Lode, west of Oak Street in Ouray.

In 1887 Weston began to spend a lot of his time in England and most of his wealth was lost six years later in the Silver Panic of 1893. The "Panic" occurred when silver prices were no longer set by the federal government. The price of silver fell drastically, and most silver mines closed. However, Weston's effort had helped to establish the San Juan mines to the point that the locals overcame these obstacles and again prospered when gold values were recognized throughout the San Juans.

A few notes about the original pamphlet: The editor has made small corrections of spelling, grammar, and punctuation without notation of such. Weston's booklet has also been made more interesting by the addition of a few historical notes (editor's notes are indicated by italics) and also by the addition of photographs (some of which were taken after the date of publication of this pamphlet). Weston's extensive assay notes may be somewhat dry for the average reader, but they have been retained as they are invaluable for the geologist or the serious mining historian.

P. David Smith

PREFACE

EXTRACT FROM THE ADDRESS OF

MR. LLOYD TEVIS

(President of Wells, Fargo & Company.)

BEFORE THE AMERICAN BANKERS' ASSOCIATION, AT NIAGARA FALLS, AUGUST 10, 1881.

"Mining for the precious metals is necessarily a business involving uncertainties. These uncertainties are of the nature of the business, and as in a lottery, it is the blanks that make the prize. If rich deposits of gold and silver could be readily and certainly found and worked continuously, these metals would lose their value. But the uncertainties which attend the search for the precious metals do not prevent the business from being a legitimate one. They merely create a peculiar need for care, forethought, knowledge and prudent energy, which gives to these qualities large rewards. Pursued judiciously, mining for the precious metals should, upon the average, return as good and sure a profit upon the capital invested as any other form of production, and there is no better and safer property than gold and silver mines, carefully selected and prudently worked.

Nor is speculation in mines, any more than speculation in other things, to be in itself condemned. It is not necessary to say to an assemblage like this, that it is a mistake to regard all speculation as evil. Without speculation there could be little industrial or social progress. Divest any community entirely of that spirit that seizes opportunities and is ready to risk where it may gain, and the energy that presses forward in new and greater enterprises would be gone. Nor would such a varying character of the seasons the alterations in trade, and the vicissitudes to which all things human are liable. For the speculator is often the economizer and provider, as he is the inventor and pathfinder. He, for instance, who, in time of plenty anticipates a time of scarcity, and buys breadstuffs for an advance, it was once the fashion to call a forestaller and monopolizer. But he is really doing for society what Pharaoh, forewarned by a dream, did for the people of Egypt, when during the seven prosperous years he stored up the surplus of abundant crops against the famine years that were to follow. Whatever enhancement of price, the operations of such a speculator may in the first place cause, will to the community at large be amply compensated for by the subsequent effect in keeping prices down. By causing economy he prevents distress; and whatever profit his forethought may bring him, is but a small part of the gain which that forethought secures to the community."

INTRODUCTION

The following pages are a revised edition of my pamphlet, written for the Denver Mining Exposition, in June, 1882, to which is added an account of the rich strikes made late last fall in the Red Mountain district, together with a general review of the past season, in Ouray County.

W. WESTON

SAN JUAN GOLD AND SILVER MINES

THE TOWN OF OURAY

(From The Denver Republican.)

Now that the Denver and Rio Grande Railway is rapidly approaching this lovely little mountain town, and its beauties will be made accessible to the traveling public, I wish to give you as a sort of a preface to a description of its mines, a sketch of Ouray itself and its surrounding attractions. Professor Rossiter W. Raymond spent a week in this country with me last summer, and as we rode along the northern flank of Mount Hayden, looking down upon the beautiful panorama of Ouray and the Uncompahgre Park, he remarked that the time would come when millions of dollars would be spent here, by tourists and pleasure-seekers, as soon as the railway reached us. And already this has begun to be proved by the rapidly increasing influx of visitors. *(The first D&RG train did not actually arrive until four years later in December 1887, but the flood of tourists has steadily increased over the years.)*

At this writing (January 25, 1883), *(This portion of the pamphlet was written in January, the rest as late as June of 1883)* the best way is to leave Denver by the night train leaving Denver about 7 p.m., taking a berth in the Pullman sleeping car which goes as far as Gunnison City, arriving there about noon the next day. Leaving Gunnison the train arrives at Montrose about 6 in the afternoon; leaving Montrose the next morning by Barlow & Sanderson's four-horse coaches, you arrive at Ouray between 2 and 3 o'clock p.m.

Ouray is the county seat of the county of the same name *(which today includes Ouray and San Miguel counties)* and was named after Ouray, the well-known chief of the Uncompahgre Utes. Captain Wheeler, of the United States Engineers, is my authority for the statement that "Uncompahgre" means the "Valley of Fountains," and in the town site of Ouray, and in the valleys below, there are numerous very large hot springs, which gush from the earth, and this fact is said to have given rise to the name; other savants say that Uncompahgre, in the Ute language, means "hot water;" which, however, is correct, matters but little; the origin is the same.

The town of Ouray is said to have had the first stake stuck in it, in July, 1865 *(sic, 1875),* when Gus Begole and Jack Eckles came from the Animas country down the canon on the Uncompahgre *(this would not be the present-day road from Ouray to Silverton, but rather what is now the Engineer Jeep Road),* on a prospecting tour, and struck the pretty little park where the town now stands. On the trip Gus Begole, now one of our principal merchants, discovered the "Mineral Farm," generally described here as "forty acres of ore," afterwards selling the same to the Norfolk (Va.) and Ouray Mining and Smelting Company. Two more prospectors came down shortly afterward, and while prospecting for gold placers, amused themselves by catching a few speckled trout. About 100 yards above where Canon creek and the Uncompahgre

A. W. Begole's substantial Ouray store was paid for by his profits from selling the Mineral Farm Mine. Gordon Kimball was another early day prospector who did the same thing. "Buenemann Bros" was a "flash in the pan." Twin freight wagons wait to be unloaded outside Begole's store.

Courtesy Western History Dept., Denver Public Library X-12814.

river join one another, they were thus busy with rod and line, when one of them noticed on the bank of the creek, an outcrop of quartz, stained bright blue and green with carbonates of copper, and breaking off a piece, rich gray copper ore was exposed to view. They then staked two claims, and called them, apropos of their employment at the time of their discovery, "The Trout and Fisherman," both being situated within the present town limits. Other pioneers soon followed, and the first cabin was built in October, 1875. About that time Captain Cutler found that wagons could be driven into Ouray from the lower country via the Uncompahgre valley, and "brought in two loads of provisions, returning loaded with Trout and Fisherman ore, this being the first ore shipped from Ouray."

This is about all that need be said of our early history. Since that time hundreds of phenomenally rich mines have been discovered in the surrounding mountains, and are being discovered now; but the simple fact that the nearest reduction works of any account were in Pueblo, and that the end of the railroad was, until last year, 200 miles from Ouray, has prevented the owners from making any great show as producers, it being all the owners could do to hold on to their property, until the railway should come to their relief, with cheap freights, cheap labor, cheap supplies and other facilities for working their mines to a profit. This relief is now nearly at hand, *(Weston, quite properly, always stresses the importance of the D&RG to Ouray. The railroad had only arrived in Montrose, thirty six miles away, in 1882)* and this little corner of southwestern Colorado will soon show a bullion product that will astonish the world. I know what I am talking about, for I have lived here over five years, working under ground for three years and a-half of that time. I have visited all the prominent mines, and assayed the different ores; my knowledge of assaying having been gained at what is generally supposed to be the best mining school in the world, viz: the Royal School of Mines, of London, England. I know the gold and silver is in the ores, and in the most astonishing quantities. There is a legend in the old country, that the rabbit begins the hole, then the

A NEW PROSPECT IS BEING MINED BY THE MEN IN THE BACKGROUND, WHILE THE TWO MEN IN THE FRONT ARE SEPARATING THE LESS VALUABLE ROCK FROM THE RICH ORE. THIS WAS ABOUT ALL A PROSPECTOR COULD DO TO LOWER TRANSPORTATION COSTS.

Author's Collection.

badger comes along and widens it out, and makes it comfortable and roomy, and then along comes the fox, and drives out the badger, and takes possession. The first owner of the mine may have to give up, and the second; but the stuff is there, and some one is going to get it out and make it pay largely and permanently; but of this more anon.

Approaching Ouray by the well-traveled toll-road *(Otto Mears had built a toll road from Saguache to Ouray via Lake City)* from the outer world, you pass through a level park traversed by the Uncompahgre River. It is about ten miles in length, and from a mile to a mile and a-half in width. At the lower end, close to the road, are some very large hot springs, the temperature of whose waters nearly reaches a boiling point, and are impregnated with iron, salt and lime, and the alkalis. *(This was the Orvis hot springs, which was a favorite of the Ute Indians.)* This park is bordered on the west side by a straight line of cliffs of sandstone capped with volcanic rock, gradually decreasing in height toward the north; and on the east by slopes more or less steep, from the Uncompahgre Peak group and its spurs. On either side of the road, as you drive through this park to Ouray, are flourishing farms, and the stream towards Ouray is well wooded with poplar, pines, box elder and willows. Within about two miles of Ouray, this park narrows into a magnificent gorge, bounded on each side by precipitous cliffs of sandstone of the carboniferous age, and sloping backward from the edge, dense forests of pine and quaking aspen timber, the whole crowned by serried peaks and truncated masses of gray trachyte, the summits of these peaks being from 3,000 to 4,000 feet above the valley, and from 10,000 to 14,000 feet above tidewater. From this gorge you emerge into the beautiful amphitheater in which stands the town of Ouray. The population is now estimated at about 1,000, and the rapid approach of the railway seems to have put new life and hope into everyone and everything; upwards of twenty large new buildings have been built during the past year, the first departure from the ordinary frame battlement front building, so prominent in our frontier towns, being made by Mr. D. C. Hartwell, commission merchant, who has erected a substantial brick building, with stone basement and iron roof. *(This building was used by Hartwell for an office but also contained the local newspaper).*

The town consists at present of the following places of business: One fancy goods and clothing establishment, one dry goods, five grocers and miners' supplies, two hardware, two bakers, ten saloons and billiard rooms, one dance hall, two druggists, two butchers, two shoemakers, one tailor, two milliners, one saddler, three assayers, one watchmaker and jeweler, two barber shops, two hotels, three restaurants, two blacksmiths, three churches, one bathing establishment, with hot and cold mineral springs, two newspaper and printing houses, three livery stables, one saw and planing mill, one furniture warehouse, one bank, and the usual number of necessary evils, such as doctors and lawyers. *(Weston discreetly left out several dozen houses of prostitution.)*

On the steep pine-covered hill to the west of the town, is the waterworks reservoir, fed by Oak Creek, a stream whose waters are pure and clear as crystal, and

This photo, taken about 1884 at the corner of 5th Avenue and Main Street, shows how Ouray grew in just a few short years after the Red Mountain boom. The pole in the middle of the street was for a huge American flag that was flown on the 4th of July.

Author's Collection.

never failing. The water is distributed all over town by pipes, with hydrants at each street corner for fire purposes, and smaller ones in front of the stores for private use — water being also turned on in most of the dwelling houses. This powerful head of water, together with a well-drilled hose company and complete apparatus, comprises our fire department, and we can now build in comparative security as well as get insurance. On the west side of town, also, is located another of our prominent local industries, viz.: that of the Ouray Brewing Company, owned by home capital and doing a fine business in supplying the town and outlying camps with an excellent article of beer. *(Breweries were often one of the first businesses of a new town in the San Juans.)*

In the southwest portion of the basin, in which stands the town, and where the waters of Canon Creek flow into those of the Uncompahgre, there are some lovely canons and picturesque gorges; and here, in places, where the hot springs flow down over the banks into the main stream, the rocks are covered with a perfect mat of "maiden hair" and other ferns. Some of these springs contain iron and sulphur in large quantities, but what other medicated matter I do not know, as we have never had them analyzed. A short distance from here up Canon Creek, is a large cave, just how large is not known, for it never yet has been systematically explored; but there are large chambers in it, the floor and roof covered with stalactites and stalagmites. This cave is entered from the edge of Canon Creek, and it is impossible to get into some of the chambers without ladders. *(Crystal Cave still exists but is now private property, and the stalactites, stalagmites, and crystals have been pilfered.)* The hot springs above spoken of, all of which flow into the Uncompahgre, serve to keep that river without a film of ice in the severest weather, both in town and for some distance below it; indeed, on a very cold morning in winter, I have seen a little white cloud of steam hanging over the river for over a mile below town. On the east side of Ouray, Portland Creek flows in, and from a cleft high up in the gigantic buttress of red sandstone, which flanks the northeastern portion, Cascade Creek rushes down and shoots far out into the air, ending in a lovely veil of spray. In the northern part of the town there are more hot springs, and where they flow into the river are large beds of delicious water-cress, which, owing to the warmth of the water and protection from frost by steam, remains green all through the severest winter, a luxury not to be despised at that season, when "green stuff" for food is so scarce and so desirable. *(The hot springs pool was eventually built at this spot.)* The Uncompahgre and its tributaries abound with speckled trout, and I have myself caught them in that stream within the town limits.

The wonderful contrasts of climate and color here are what make the scenery so indescribably charming. Even now (May) while I write, it is warm and summery in town, the side hills are covered with flowers and the whistle of the humming bird's wing is heard in the air; yet I can look up at White House Peak and see the snow banners blowing from its summit, as in the coldest day in winter. In the autumn, more especially, are the contrasts of color seen, and the landscape, as it then appears, if painted on canvas, would, I believe, be laughed at, if shown in Europe or the Eastern States, as impossibility. I have been climbing the heights above Ouray, and looked

Even the early prospectors and miners took time to enjoy the magnificent natural beauty of the San Juans. This photograph was taken near Imogene Basin about 1900. The same beauty is enjoyed today by hundreds of thousands of tourists and natives.

Author's Collection.

down on it, when the atmosphere of the valley seemed of a hazy blue, the sloping sides of the surrounding mountains being clothed with the golden yellow and the red brown of the quaking aspen and the dwarf oak, the varied greens of the spruce, balsam, cedar and yellow pine, and above the brown gray of the trachyte peaks, their snow-capped summits forming a charming contrast against the lovely violet and blue of the evening sky. Such colors in a landscape would seem unnatural to eyes unused to mountain scenery.

Ouray is situated at the head of

UNCOMPAHGRE PARK

In a sort of basin, or amphitheater, where the waters of Canon Creek and Uncompahgre join, and thence flow downward through the valley. All around the town, the mountains rise to altitudes of 13,000 to 14,000 feet, presenting, above timberline, a magnificent array of cliffs and peaks. The streams which, coursing down from these mountains, swell the waters of the Uncompahgre, are Sneffels and Imogene Creeks, which, at their junction form Canon Creek, and have their source at an elevation of about 12,000 feet above tide-water; Red Mountain Creek, draining Red Mountain Park at an elevation of 2,000 feet above Ouray, and Bear Creek, which, where it flows into the Uncompahgre, about two miles above Ouray, has a sheer fall of over 200 feet. All these are large and never-failing streams.

Professor Hayden, in his Colorado survey reports for 1879 *(his 1873-76 reports were published in 1879)* says: "The Uncompahgre River, after leaving the San Juan mountains, where it heads, flows northward through

Uncompahgre Park. This is a small but fine valley ten miles long by two wide. At the foot of the park a large branch known as Dallas Fork enters the Uncompahgre from the west. Fine meadows extend up this fork for four or five miles." If Professor Hayden could visit the valley of Dallas Fork now, he would find his "fine meadows" turned into flourishing farms, and large herds of cattle grazing on the luxuriant grasses of the mesas on either side. He also says, speaking of the park: "Its low elevation, compared with most of the valleys of Colorado, gives it a climate sufficiently warm for the production of all kinds of grains and garden vegetables," etc.

THIS RECENT PHOTOGRAPH SHOWS THE UNCOMPAHGRE VALLEY IN THE WINTER. THE VALLEY IS JUST LOW ENOUGH IN ELEVATION THAT THERE IS OFTEN SNOW IN THE MOUNTAINS WHILE IT IS RAINING IN THE VALLEY — A PERFECT SITUATION FOR FARMS AND RANCHES.

Author's Collection.

(Hayden, a geologist, did his survey to determine if the San Juan mountains contained enough precious minerals to force the Utes out of this part of their territory.)

This has been proved by the farmers and market gardeners in the park, who are now successfully growing grain, and vegetables such as cabbage, onions, beets, carrots, tomatoes, potatoes, turnips, and lettuce. Strawberries are being extensively and successfully grown this summer. Wild strawberries abound, and there can be little doubt as to the successful cultivation of bush and tree fruits, as black currants, gooseberry, raspberry, cherries, squaw berries, and Oregon grapes grow wild here in profusion. Mushrooms in large quantities, and truffles are found about Ouray and in the park, Professor Hayden also says, comparing it with the lower end of the Uncompahgre Valley, and showing how the difference in elevation influences the character of the vegetation: "In the interior, near the head, the land is to the Utes, one flowing with milk and honey. Here are his streams of pure cold water, beautiful aspen groves, the best of the grass in the greatest abundance, and a profusion of wild fruits and berries, while the country is a perfect flower garden. This extends as low as 7,000 feet, below which the scene changes to one in all respects the reverse. Aspen gives place to pinon and cedar. The grasses, fruits, and flowers, to sage, cacti, and bare rock. The streams become confined in rocky canyons, turn muddy and narrow, and gradually disappear."

It may be well to mention in this connection that the Ute Indians have all been moved away, for which we are all most truly thankful, as they are a curse to any country. *(As a result of the Meeker massacre the Utes were removed from Colorado in September 1881, which opened up the rest of present Colorado north of the San Juan mountains.)*

Farming in the park has for many years proved a most lucrative business, and as the producing area here is limited, and the consuming area — i.e. the mines — boundless and ever increasing, there will always be a ready market for everything raised, though of course the completion of the Denver and Rio Grande Railway to Salt Lake and the Pacific coast will reduce prices very materially in some articles of consumption, and our flour and dried fruits, which we now get from the other side the "Great Divide," will come from our neighbor, Utah.

GAME

In the autumn, when the water fowl are leaving their summer habitat and migrating southward, they pass over the Uncompahgre Park in numberless flocks, at a height in the air of from one to two thousand feet, following however, as to their route, the course of the main stream. But when they come to the divide, or crest of these mountains, they have to elevate their flight, say fifteen hundred to two thousand feet in order to cross over. This happens, at any rate in cloudy weather, for I have often watched them. In fine weather they may, for all I know, pass over far out of sight of human eyes. However, sometimes they strike this valley during the equinoctial gales and in a snow storm, and flying straight up the Uncompahgre Gorge, they soon find the mountains closing in on every side; the sky being dark and the air thick with snow, and no outlet to be seen, they become bewildered, and the large flocks of ducks and geese and other wild fowl return to the valley, where they stay about the beaver dams and pools of the Uncompahgre River till the sky clears and they can continue their journey toward the Sunny South. I have seen, during one of these snow storms, flocks of ducks and geese flying about over the town of Ouray, and at that time hundreds were killed in the park, one man alone killing over a hundred.

In the fall of 1881, my partner and I had a few days' shooting in the park, and I kept a list of varieties that we killed, and of others that I saw and did not kill. They are as follows:

Anatidoe-Mallard, Gadwall, American Widgeon of Bald-pate, Wood-duck, Spoon-bill or Shoveler, Green-winged Teal, Blue-winged Teal, Ruddy Duck, Hooded Merganser, Sheldrake, Gooseander, Pintail, Canvassback, Red-head or Pochard, Blue-bill or Scaup Duck, Buffel-head, Canada Geese, and Brant.

Gruidoe-Sand-hill Crane.

Ardeidoe-Great Blue Crane, Brown Bittern.

Falconidoe-Osprey, Chicken hawk, Sparrow hawk, Falcon.

In addition to these, which are not worth classifying, were Yellow-leg plover, Kildeer plover, Avocet, English snipe, Grouse, Jack-rabbit, Common rabbit or "Cotton-tail," Crow, Raven, Magpie, Doves, Meadow-lark, Yellow-hammers, Red-winged Starling, Sandpiper, Red Breasted thrush, White-rumped shrike, Cat-bit or "Camp-robber," Crested Jay, King-fisher, Blue bird, Swallow, Night-hawks, and Humming birds.

Of the water fowl I had killed all the species, both on the Canada lakes and in the marshes of Missouri and Kansas, excepting however, the Avocet, of which we bagged nine, and which were the first I had seen except in collections. They belong to the Recurvirostridoe, their long bills turning upward, instead of being curved downward like the Curlew.

Two years ago a fine specimen of the Loon, or Great Northern Diver, was caught in one of the irrigating ditches in the park and sent up to me, and I have him now, stuffed and mounted. And this spring a fine specimen of the Eared or Crested Grebe was sent to me, which I have also preserved. How these two birds strayed from the big lakes they usually inhabit, into a little mountain park, is hard to tell.

But I was more than ever astonished to meet at this altitude, my old friend, "Scolopax Wilsonii," or English Snipe. It was one day in November, during a heavy snow storm, it having already fallen to the depth of a foot in town, that I walked down to the hot springs in the North End to get some water-cress, and found that for a few feet round them the mud was bare of snow, owing to the heat and steam. To my surprise, up jumped a wisp of eight or ten snipe, and flew up stream to some more hot springs where they pitched. I immediately went and asked leave of the town marshal (as there is an ordinance against discharging firearms within town limits), and getting it, went back with my gun and bagged three couple of the long-bills. Last January I got three more on the same place. During the same snowstorm spoken of above, a poor bittern, tired out and bewildered by the snow, settled in the main street of town, and was captured alive. *(An amazing variety of birds, but Weston was not exactly a conservationist as we know one today.)*

With big game this district abounds. There are beaver in the park, and in the hills around it are black-tailed deer, mountain sheep (ovis Montana), cinnamon, black, and frost bear, coyotes and foxes. In spring and fall the hills and mesas surrounding the park are full of deer, but in the middle of the summer they find their way up into the high mountain parks, where they can graze during the day in the succulent bunch grasses above timberline, and there be free from annoyance by the flies. Of the game in the higher altitudes, however, I shall have more to say when I come to my description of the mines.

On the north side, and where you emerge from the gorge, into the basin where stands the town, are almost perpendicular bluffs of red sandstone of the Carboniferous age. These partially encircle the town, while round the other portion the hills are more sloping, and covered with pine timber, quaking asp, and shrubs of various kinds. I estimate the average height — or rather thickness — of this sedimentary formation, to be eight hundred feet. Above this is the layer of trachyte or volcanic rock, of an average thickness in this district of about three thousand five hundred feet.

On the south of town is a stratum of sandstone above the limestone, and the southwestern portion of the town have for their walls so-called quartzite, or really altered sandstone, their ores being contained in a gangue or matrix of quartz porphyry. Above the Ophir Mine is a stratum of "ribbon jasper" of beautifully variegated colors, and here also I find quartz conglomerates in the shape of both pudding-stone and breccia. The latter, with its contained angular fragments of varied colors, would look very handsome in polished slabs for tabletops, etc. The effect of heat in metamorphosing the sedimentary rocks is here shown in a very marked manner.

The steep cliffs around Ouray close in on the town on all sides except where the river flows out of town. Ouray's streets were still under construction in the early 1880s. A landowner could physically work on the streets to pay his real estate property taxes.

Author's Collection.

Leaving Ouray at the south side, you ascend the picturesque and heavily wooded gorge of Canon Creek, towards Imogene, Sneffels, and Virginius Basins, in which are located most of the celebrated mines of the Sneffels' district. About three and one-half miles from town, and one thousand feet above it, you see the last of the red sandstone where the creek has cut through it, and you are now between walls of trachyte, the sheer bluffs on the east side stretching upward in an almost perpendicular line for over a thousand feet. The west side, however, is more broken and sloping, and here the Sneffels toll-road has been constructed. *(Although the county started this road, Otto Mears was called in to finish it.)* From this point of junction of the sedimentary and igneous rock, to the summits of the highest peaks, the trachyte prevails.

Professor Hayden has classed this trachyte as No. 4, or the youngest of the four groups of trachyte rock. He says: "this group is, horizontally, a restricted one, but fraught with occurrences of the highest interest. Not only are the rocks themselves of very peculiar type for the position they occupy, but the presence of many metallogerous veins lends additional importance to the group. It occurs in the heart of the Uncompahgre group and extends to its western edge. Some of the highest mountains of that region are partly or wholly composed of it. The latest discoveries of ore-bearing veins seem to have been made at localities where this group occurs as merely capping the older ones. In other words, instead of the veins being confined to No. 4, they extend through it, and can be reached in older formations." Speaking of these lodes about Silverton, he says: "Nearly all the lodes located near Baker's Park were found in trachyte No. 4. Subsequent investigation has shown that they penetrate beyond the limits of this group, and without an appreciable change of course or character, enter the metamorphics, which are covered by the trachytes. All the veins that I had occasion to visit, located within the trachyte, were argentiferous. The only one which was worked at that time (1876) in the underlying schists, the Little Giant, is auriferous. I am not aware if this feature has since been established as the rule." It has not. For just over the divide from the Sneffels district, in the San Miguel district, the Smuggler, Cimarron, Argentine, Cincinnati, and other veins, which are gold as well as silver bearing, are being worked in the trachyte formation. There is also a lode, the Governor, bearing free gold in Sneffels, in trachyte.

Stating the general mineralogical character of the rocks of this group, Professor Hayden says: "The paste is compact — rarely vesicular — and microcrystalline. Oglioclase is the most prominent of the enclosed minerals. Another feldspar — triclinic, probably andesite — occurs with it. Sandite may be regarded as an exception, and occurs only in the lower members of the group. Pyrite, magnetite, hornblende, mica and chlorite are accessory minerals, of which the first named occurs in the shape of an impregnation."

On the east of Imogene Basin, and southeast from Ouray, is Red Mountain Park. To reach this from Ouray, you take the toll-road *(a county toll road, as Mears had not yet started the Million Dollar Highway on the east bank all the way to the top of Red Mountain)* to a short distance above Bear Creek falls, and there crossing the Uncompahgre, follow the Red Mountain trail. This trail leads to the head of the park, and you can then, by taking the left hand trail by Grey Copper Falls, reach the head of Cement Creek, and thence down to Silverton *(the present Corkscrew Gulch route)* or you cross the divide between it and Mineral Creek, and then follow the course of Mineral Creek to Silverton *(the present U.S 550 route),* the distance either way being about 22 miles.

In this Red Mountain Park, at the upper end, are some brilliant scarlet peaks, and as there has been some difference of opinion among our local savants as to the origin of their color, I will give Professor Hayden's description. Speaking of No. 4 group of trachyte, he says: "On the point of petrographic character, this group presents more attractions than any of the volcanics. Many brilliant colors of the most delicate shades are shown by the mountains and ridges. They are due to admixtures of certain mineral substances. Dark colors may be said to be characteristic of the main bulk of this group, but a very prominant exception is made by what I have termed the 'red stratum.' Originally white, the presence of ferric oxygen compounds gradually changes this color to yellow, orange, red, and brown. The rock is a microcrystalline feldsparic paste of white color, containing very minute transparent crystals of sandite and small crystals of pyrite. Throughout the district, wherever this stratum could be traced, the crystals of pyrite were contained in it as an impregnation. Decomposition of pyrite releases the sulphur and changes the iron from a bisulphide to hydrated sesqui oxide. This, in varying percentages, produces the colors and shades above enumerated."

On Bear Creek, from two to four miles from Ouray, a good many silver-bearing lodes are found and worked in the quartzite formation, some of them extending

Bear Creek ends in a beautiful waterfall that drops under the highway and into the Uncompahgre River. Otto Mears chose this point for the tollgate for his toll road, as there was no possible way to get around it.

Author's Collection.

upward into the trachyte without, apparently, any material change.

In writing of our mines I must need to do so in a very general sort of way, as were I to particularize each lode claim that I have seen here, it would fill an extremely big book. My chief object now is to give a general idea of our ores and the facilities for working them, and in doing this, I can only describe those mines which have had some substantial development work done — those which I have personally inspected, and which will be fair examples of their class. I have in a former letter described the geological formation of each basin or gulch where the principal mines contiguous to Ouray are located.

THIS SCENE FROM A *HARPER'S WEEKLY* ARTICLE WAS DRAWN FOR INVESTORS AND POTENTIAL PROSPECTORS BACK EAST, AS MINING CLAIMS WERE 300 FEET WIDE AND 1500 FEET LONG. IT IS ENTITLED "GOLD AND SILVER MINING IN COLORADO — A HONEY COMBED MOUNTAIN."

Author's Collection.

To begin with, there are the mines below town and in town, so to speak. One of the most remarkable of the former is the Dexter, about three miles from Ouray. It is situated on the east side of the park, and not over a mile from the toll-road *(another Mears' toll road, this one from Ouray to the Los Pinos II Indian Agency to Lake City.)* and I may here remark parenthetically, that this toll-road is the line already surveyed for the Denver and Rio Grande railway. The Dexter is in a shale and quartzite formation, and is more of a horizontal than a vertical fissure, dipping into the mountains at an angle of about 33 degrees. The vein matter is porphyry, and the ore galena, antimonial galena, blende, and in seams on the galena are large quantities of native silver in the form of wire and "moss" silver. About 500 feet of work has been done on the vein, and nearly all of it in ore. Prior to its purchase from the original owners, the ore was gouged out by them, and the mine badly treated, but even at that three tons are known to have been sold, which were milled $1,500 in silver to the ton, and 500 pounds were sold, which realized $700. However, my estimate of the average of this ore is 50 ounces silver to the ton and 20 per cent lead, the width of the ore-streak being eighteen inches. I am informed that there are 300 tons of ore on the dump, and that what has been taken out on doing the dead work will more than pay for the same.

Within half a mile of town and opposite the tollgate, is the mine known as the Grand View. This lode breaks up through the red sandstone formation, but has not been found, or even prospected for, in the trachyte above, as for a very long distance above and beyond the sandstone, the mountain is gently sloping and covered with a deep alluvial deposit and a dense growth of pine and poplar timber. Looking at this mine from the road, the sandstone cliff had the appearance of a "broken back," as we call it; that is, when the fissure was formed the stratification was broken, and the rock tilted up the almost perpendicular bluff, and can be worked to a depth of at least 800 feet below the surface by driving levels on the vein alone. The ore is iron and copper pyrites, with the results of decomposition of the latter — viz. blue and green carbonates of copper and occasionally a little galena scattered through it, the gangue being quartz porphyry.

In the upper level they had ore from the entrance to the breast (about 160 feet, I believe) and the streak varied in width from one to three feet. I once made a test of this ore, from entrance to breast, and the result I got was an average width of two feet of ore that yielded $60 in gold and $10 in silver to the ton of 2,000 pounds, and in making this sample test, I took such a very liberal amount of poor rock, that I hardly think it did the mine justice; still I maintain that my results were "good enough." However, mill runs of the ore taken from this level — no stoping done — have yielded $100 to $150 to the ton, $14 to $20 being silver, the rest gold; while assays of samples have given as high as 43 ounces of silver to the ton, and 25 ounces of gold. No free gold has been found that I am aware of, and it is contained, I believe, in the pyrites. I am convinced that this vein is a true fissure (although some doubt has been cast on it, but the workmen are entirely to blame for this in running off the vein in the lower level) and carries ore so similar, that beyond doubt the ores and vein matter of the two have a common origin. *(Weston keeps mentioning fissure veins, as they could run deep into the earth rather than appearing in pockets.)*

The Grand View Mine was located on the cliffs to the right. The Great Western Mine was at the bottom of the photo. It was connected by tram to its mill on the valley floor. The City of Ouray is in the background with the Otto Mears toll road entering the town from the bottom of the photo.
Courtesy Western History Dept., Denver Public Library F37534.

Across the canon, and close to the roadside, is the Great Western, also on sandstone formation. At the lower end it is apparently some forty feet from wall to wall, the ore being in separate streaks at about a foot wide, and divided by horses of country rock or altered sandstone. This is on the outcrop. A level has been run on one of the streaks, and large quantities of massive copper and iron pyrites taken out, some of which ran well in gold, with a high percentage of copper. The ore streaks seem to come together as the vein ascends the mountain, which is so acclivitous, however, as to render an examination somewhat difficult. I am of the opinion that the Great Western is on the same vein as the Grand View, although many differ with me. *(Many later geologists supported the same view. Placer gold found near Ridgway probably was washed downstream from this spot.)*

Above this mine, and about 900 feet vertically above the road, is an immense flat or horizontal iron lode, at the contact between the sedimentary and volcanic rocks. It outcrops for over 3,000 feet, and varies in width from three to nine feet of solid ore. In some places there are three feet of red hematite, and in others it changes to massive copper and iron pyrites; the hematite ore, however, prevails. In one place there are large vughs, or cavities on the ore, their interior surface being encrusted with sulphate of alumina. The hematite was used as a flux by the Norfolk and Ouray smelter, which is located on the river immediately below the two claims, and it carried uniformly $10 to the ton in gold and silver. A piece of ore can be dropped from these two mines onto the toll-road. Assessment work has so far been done only, and on the outcrop of course, and many think that rich ore could be developed by running in on this enormous ore body. There are two claims on this lode, known as the Ironclad and Finance. A good trail has been built from the toll-road, and you can ride a horse up to the mine without getting out of the saddle. In connection with this iron lode, I may state here that on Carbon Creek *(now called Coal Creek)*, about four miles from town there is an outcrop of abut four feet of bituminous coal, and about 12 miles below town a vein of very clean, bright coal is being worked by the owners and delivered in Ouray at about $14 per ton, for domestic and blacksmithing purposes. In fact, from 12 to 15 miles from Ouray, the whole country seems to be under laid with coal, and in one place, about 35 miles distant, it shows on its outcrop on the hillside a vein 30 feet in thickness.

There is near the Dexter a remarkable lode, the Calliope, which carries native silver in its quartz, and mixed up through the quartz are pieces of jet. It is a true fissure vein, and is thought by some to extend downward through a coal measure, the coal being brought up and mixed among the quartz in the process of vein filling from below. But it is the only instance I ever heard of or saw, of jet in silver-bearing quartz.

On the south of town, and on the bank of the river, are two mines known as the Trout and Fisherman, mentioned in a former letter. They intersect one another in the river bed, and were owned by the same men; in brief their history is this: As nearly as I can ascertain, ore to the value of $20,000 has been taken out of them, with work which I should estimate as costing about $5,000. The owners did not work it themselves, but leased it first to one and then another, who took out the rich ore when they came to a "fat" place — for the ore is "bunchy" — and then threw up the lease as soon as they came to a lean place. In 1876 ore was taken from the outcrop of the Fisherman by leasers, which milled, first-class, 804 ounces of silver to the ton, and second class, 300 to 450 ounces. The ore is gray copper, copper pyrites, iron pyrites, a little galena scattered through, and the blue and green carbonates of copper, the matrix of the metals being quartz-porphyry and the wall rock altered sandstone. The gray copper is very highly argentiferous, and this ore has been sorted to run 600 ounces silver to the ton, while specimen assays of the gray copper run from 1,500 to 2,000 ounces to the ton. *(Although the Trout and Fisherman looked promising, its rich ore was basically only on the surface.)*

A few hundred feet higher up the creek is the Ophir. This mine is in a similar formation and has the same ore in the same gangue as the other two, but the vein is more clearly denned. The tunnel was commenced just above the level of the river and run in 100 feet, with a shaft sunk 25 feet. The ore shows an average width of 15 inches and will average at the mill about 150 ounces of silver to the ton. Selected lots of first-class ore from this mine have yielded as high as 1,500 to 2,000 ounces silver to the ton, and a lot of second-class that was hand-sorted — copper pyrites and galena — yielded 368 ounces silver to the ton. The shaft is all in ore and the width of the ore streak in the shaft is two feet. These last three mines are all worked from the riverbank, and will need small hoisting engines before they can be systematically explored.

Half a mile above these is located the renowned Mineral Farm, mentioned in a former letter. This property was taken up as four regular parallel claims, 1,500 by 300 feet each above the town, and in pine and poplar timber. Some contend that they are four parallel vertical fissure veins. My opinion is that it is a deposit of ore on a limestone bed; similar to those in the Banat, Hungary, and that they extend more or less over the greater part of the 40 acres.

At any rate, there has been enough ore exposed and deep enough already to make it practically inexhaustible; what has been taken out so far has been simply quarried from the surface, and there seem to be literally acres of it.

The ore is of different kinds. One is a bright, fine galena, running about 1230 ounces of silver to the ton, and 40 per cent lead; another is silver fahlerz, or highly argentiferous gray copper, in a gangue of baryta of heavy spar. This ore has milled from 400 to 700 ounces of silver to the ton, and assays have been had as high as 3,000 ounces. There is also a low-grade galena, carrying a good deal of antimony and about 30 ounces silver to the ton. The matrix of the metals is of two kinds — the one a quartz porphyry, and the other, as before stated, baryta.

Bernhard Von Cotta's theory for the similar formation in Hungary is that the minerals were first deposited by thermal springs, on a limestone floor, the limestone having the effect of precipitating the minerals then in solution, and that the minerals so precipitated were crystallized by the volcanic or lava rock that subsequently over flowed them. It is true that there is now no volcanic rock on the top of the Mineral Farm ore; but I think that there has been, and that it has been worn off by erosion, as large boulders of trachytic porphyry have been found on the ground. *(The Mineral Farm was worked until the 1930s and produced over a million dollars in ore, worth ten times that amount in today's values.)*

Having described some of the principal mines "in town," so to speak, I will now treat those up Canon Creek and in the Sneffels District, compromising Imogene, Sneffels and Virginius Basins.

After leaving the Mineral Farm, going southward along the Mount Sneffels toll-road, the next known mines of any importance are about two miles from Ouray. At this point, on the west side, is a deep, heavily-wooded gulch, the bed or bottom of which is traversed by an immense

The Mineral Farm Mine quickly became one of Ouray County's largest mines. It had plenty of ore that ran in trenches, which were easily mined. It was also only one mile from the City of Ouray. All of these factors led to it being very profitable.

Courtesy Western History Dept., Denver Public Library

vein, three claims on which are known as the Malta, Japan and Moslem. This property is owned by a single individual, and is 4,500 feet in length by 300 in width, and located in dense pine and quaking-aspen timber. The vein has a course nearly northwest and southeast, and outcrops boldly on the surface for nearly the whole length of the Japan and Moslem as it cuts through the mountain. The altitude of the lower end of the property is 8,500 feet above sea level, and the upper end 10,650, so that a tunnel driven at the lower end would gain a depth, by drifting on the vein alone, of 2,150 feet below the surface. This lode is a perfect fissure, 12 feet between the walls, in trachyte formation, has soft porphyry for the vein matter with a clay gouge or selvage, and the ore is of two characters — a fine grain galena, carrying 20 to 24 ounces of silver to the ton, and 60 to70 per cent lead, this ore streak varying in width from one to four feet; heavily mineralized with the same fine galena, and carrying brittle and native silver scattered through it. The great size of the ore body and vein has rendered it impossible for the owner of limited means to prospect it properly and determine how good this latter streak is, but it gives every promise of an enormous body of very rich ore, and with the adjoining continuous body of lead it is certainly a premium smelting ore. A tunnel has been driven on the Japan 37 feet, and on the Moslem 75 feet, and these, with several open cuts, are all in the ore. There are sites for at least 15 drifts on this vein, and the ore can be extracted at an extremely low rate of cost. *(As good as Weston makes then sound, these claims were never developed to any great extent.)*

Going on up the toll-road, at a distance of about six and a half miles from Ouray you come to the foot of

What came to be known as the Camp Bird Road was a busy route filled with burros, mules, and freight or ore wagons — all coming and going to the mines. Burros were driven from behind, like sheep; while mules were led and tied together with ropes. Either way there was plenty of dust!

Author's Collection.

Imogene Basin, and the forks of the road; for here the roads turnoff toward Sneffles and Virginius Basins on the west, with one branch, while the other continues half a mile further to the center of Imogene Basin. This basin is crossed at right angles by eight well-defined lodes, and they may be seen on the surface of the opposite ridge, or west, as they cross over into Sneffels Basin, and in some instances, the porphyry vein matter and ore bearing quartz, being decomposed more readily than the trachyte, which they intersect, have cut deep sags, or notches, in the summit of the ridges where they cross. The average width of these lodes from wall to wall is about 5 feet, their strike or course being about northwest-southwest, converging slightly towards the east, or Red Mountain Valley, in the direction of the scarlet peaks mentioned in a former letter, and which we have always supposed to be our volcanic center.

(Original editor's note — Since writing this in June '82, wonderful discoveries have been made at the base of this range of Scarlet peaks, particulars of which are given at the latter end of this pamphlet).

From the great uniformity of these lodes per se, I regard the whole group as a system of fissures formed contemporaneously: and from the great similarity of the character to the ore, I think the ores and the vein matter they contain had a common origin. They were prospected as early as '75, and many claims staked; but in most cases only those were retained which showed rich surface ore, and these were eroded to a considerable depth by glacial and other action, and of a consequence most of them are above timber line or nearly so. In those days prospects could be sold on rich specimens, and the main object, therefore, was to get a claim off which you could knock a rich specimen, no matter if inaccessible — owing to snow — for eight months out of the twelve. These same veins, where they cropped out down in the valley or gulch, showed splendid bodies of ore, but on assay proved low grade, and the prospectors of that day made up their minds that these lodes were "filled in from the top," and that the only rich mineral they contained was that found in them where they crossed the highest peaks. When my partner and I prospected these lodes in the spring of 1877, and looked at nature's vast dumps — acres upon acres of "slide-rock" from 25 to 200 feet in depth, lying at the foot of the mountains in the basin — it occurred to us that, after all, nature had done the mining and exposed rich ore in the veins on the present summit of these worn down peaks, and that claims down in the gulch on the same veins, when depth would be gained on them by drifting, would show equally rich ore as that exposed in the highest summits; and within the last year our theory has been proved correct, notably in the case of the Yellow Rose, hereinafter described. *(Actually, most ore in the San Juans is richer the higher the elevation.)* The existence of such a thing as gossen, iron hat, or the altered outcrop of lodes does not seem to have occurred to the early day prospector; a lode bore just such ore as he was able to knock off the top; and whether the top then existing was the original apex of the lode he didn't stop to think either. These Imogene Basin lodes dip toward the north and have a strike about northwest-southeast. The ores have white crystalline quartz as the matrix of the metals, and galena as their base, including highly argentiferous gray copper or silver polybasite, silver glance, native silver in leaf and resin blende, iron pyrites, peacock copper, and the blue green carbonates of copper. Calc spar occurs occasionally, both in the form of "dog-tooth" and "nail-head," and in Yellow Rose vein, large quantities of the silicate of manganese, and also the pink spar of carbonate of manganese, the latter being almost always accompanied by native silver and silver glance. These ores are usually of a "combed" or "banded" texture, the layers being parallel to the fissure. The quartz and ore is generally on the hanging wall, with a clay gouge or selvage between the former and being, as before stated, trachyte. This description, of course, only applies to the veins when "in shape," as like all lodes they are subject to "caps," "pinches," and other changes.

As an example of what development will do in these veins, I will cite one interesting instance — that of the Yellow Rose. *(Weston goes into great detail on the Yellow Rose and other Imogene Basin claims, as most were developed for several years by him and his partner.)* This vein is a strong true fissure, has 10 locations on it, and can be traced on the surface for four miles, and wherever it crosses the highest ridge of mountains shows rich ore on the surface. Where my partner and I commenced work on it in 1877 we found it pinched to about a foot, the average width of the vein being about 7 feet. We drove 60 feet in it without a sign of ore, and we then got a good streak of ore — which afterwards proved to be a pocket — and sold with other claims to a company who drove about 150 feet further without finding ore of any importance. For the next 200 feet they had ore from 1 to 2 feet in width, running from 20 to 130 ounces silver to the ton, the ore being principally bright galena, associated with iron and copper pyrites, gray copper, and blende, in a gangue of white crystalline and very friable quartz. Then a sample assay of about 8 inches of galena gave returns of $325 to the ton, which somewhat astonished the Superintendent. But a few more shots in the ore threw out large masses full of pockets of native silver glance with wires of native silver sprouting out of them. In 50 feet more this ore body widened out to 3 feet, which is the width of the ore in the breast at this writing. These are not occasional pockets of silver, for in a breast of ore 3 feet wide pure metal seems to be evenly distributed through it, as it is found in almost every mass thrown by the blasts. The native silver in this vein is almost always associated with pink manganese spar, and in Sneffels lodes this has so far been nearly always an indication of very rich ore. The foregoing rich strike was made 480 feet from the ground of the Declaration, the adjoining claim to the Yellow Rose, as the owners of the latter ran their drift 42 feet beyond their end lines. The Declaration and its adjoining claim, the Agnes, can be worked to a depth of 1,000 feet below the surface by drifting in the vein alone, and every miner knows how comparatively cheaply ore can be extracted when you can drain your mine and run out your ore by levels without needing machinery to pump and hoist.

The fact is the ore at all of our best veins runs in chutes or chimneys; and between these are barren or lean places, and you must gain depth before you can get rich ore in this district. Of course rich ore is occasionally found on surface here, but is generally where the vein has been worn down and subjected to Nature's mining. *(The Yellow Rose was later consolidated into and became an important part of the famous Camp Bird Mine.)*

Two claims adjoining the Yellow Rose and Declaration, on a parallel vein, are the Hoosier Girl and Richmond. They are full claims, located in heavy timber, and with a mill site of five acres, also heavily timbered. Very little systematic exploration has been done due to the fact that the owners cannot afford it, but a tunnel of 100 feet, and some open cuts, together with where the lode has been laid bare and cut down by the action of Richmond Creek, show some remarkably pretty and rich ore, which is essentially adapted to the concentration process. The gangue of the ore is white and also a blue crystalline quartz, very light and also friable. This quartz carries brittle silver, highly argentiferous gray copper and somewhat of zinc blende, copper pyrites, and a sprinkling of galena. Some lots of first-class ore from this property milled 165 ounces silver to the ton, and a lot of second-class 65 ounces to the ton, and as it will concentrate at least six tons into one, I believe this property will pay largely when properly worked. The title is a United States patent.

The Millionaire, an extension of the Yellow Rose, is a good claim, but undeveloped.

The Sneffels toll-road runs to the foot of these five last named claims, and so does Imogene Creek, with a fall of 100 feet close to the dump, the water being never failing the whole year round. The lodes are also in heavy pine timber for their entire length, and can be worked the year round with ease and comfort. The Hidden Treasure is another representative lode of Imogene Basin, and a remarkably fine one. In the upper level, the first one run, there is a chimney of ore over 100 feet in length, which varies in width from 1 foot to 18 inches, and carries from 95 to 250 ounces silver to the ton. In a lower drift, which gains as far as driven, but at little depth below the surface, the ore was large cube galena, associated with red or "resin" blende, this ore carrying from 20 to 40 ounces of silver to the ton, the chimney being about 70 feet in length and from 8 inches to a foot in width; and after

The Hidden Treasure Mine (shown here in 1925) was a part of the Camp Bird Mine, but Walsh thought so much of it that he kept it when he sold the rest of the mine. He thought the Hidden Treasure still had great hidden riches and told his family to hang on to it. They still own it to this day.

Author's Collection.

running about 30 feet in lean ore, another chimney was encountered of a higher grade, and running from 30 to 160 ounces, the streak widening to 3 feet. This chute, however, is a continuation downward of the one encountered in the upper drift. A mill-run of some of this ore gave returns of 101 ounces of silver per ton and 42 per cent. of lead. The character of this ore is galena and gray copper, associated with iron and copper pyrites, and red blende in a gangue of white, "sugary" quartz. About 500 feet of work only has been done in the Hidden Treasure, 150 feet of it being a cross-cut, and yet ore chutes of a length of 220 feet have been uncovered. Both its extensions, the First National and Pocahontas, show as large a body of the same ore, which gives an idea of the permanence of the ore bodies in these veins.

The Hancock and Tuscola are remarkable claims on a adjoining vein, and show for nearly their entire length, on the outcrop, from 1 to 2 feet of copper and lead ore running well in silver. The other lodes of this basin are of similar character, but have not had enough development done to prove them.

Between Imogene Basin and Sneffel's Basin is a spur known as Ruby Mountain, and on this is situated the celebrated Wheel of Fortune group of lodes.

The Wheel of Fortune is a cross lode to those of this belt, having a north and south course, and carries ore of a different character from the rest. It is in trachyte formation and is about 7 feet between the walls, has soft porphyry for vein matter, and very light, white quartz as the gangue of the ore, the character of the latter being brittle silver, occasional ruby silver, and gray copper, with the blue and green carbonates of copper. Three short levels were driven on the vein by the original owners and discoverers, and as Mr. Moritz Stockder, M.E., of Lake City made a careful report on the property in the interest of English capitalists, I give the following extract, being the most direct evidence I can adduce as to the quality of the ore. *(The Hidden Treasure was a very successful mine, being eventually incorporated into the large Revenue Mine at Sneffles.)*:

MILL RUNS

"From Mr. Bouton, assayer and analytical chemist, of Lake City, Colorado, I received a statement of the following lots of Wheel of Fortune ore which were sold to the lixiviation works of that price;

July 13, 1877.	2,144 pounds.	196 ounces silver per ton.
July 13, 1877.	10,688 "	205 " "
July 28, 1877.	17,455 "	201 " "
July 28, 1877.	2,019 "	723 " "
July 30, 1877.	3,200 "	260 " "
July 30, 1877.	4,500 "	170 " "
Aug. 2, 1877.	15,300 "	212 " "
Aug. 9, 1877.	7,340 "	192 " "
Aug.22, 1877.	3,737 "	196 " "
Aug.27, 1877.	7,175 "	205 " "
Sept.13, 1877.	4,828 "	200 " "
Oct. 4, 1877.	7,370 "	160 " "

From Munn Brothers' assay office and sampling works at Ouray, Colorado, I also received certificates of the following lots of Wheel of Fortune ore:

July 22, 1876.	1,100 pounds.	640.46 ounces silver per ton.
July 30, 1879.	3,422 "	168.61 " "

July 30, 1879.	498	"	884.50	"	"
Sept.17, 1879	3,582	"	155.45	"	"
Sept.25, 1880	2,753	"	376.59	"	"
Oct. 9, 1880	410	"	290.10	"	"
Oct. 9, 1880	67	"	1,002.70	"	"

This makes a total of 97,581 pounds, with an average value of 221 ounces of silver per ton.

Messrs. Munn Brothers' certificates also show the gold value of the Wheel of Fortune ores assayed by them as follows:

July 30, 1878.	3,422	pounds.	0.38 ounces gold per ton.		
July 30, 1878.	498	"	1.90	"	"
Sept.18, 1879	3,582	"	0.31	"	"
Sept.25, 1880	2,753	"	0.29	"	"
Oct. 9, 1880	410	"	0.38	"	"
Oct. 9, 1880	67	"	0.50	"	"

This makes a total of 10,732 pounds, with an average value of 0.43 ounces of gold per ton."

After the property was sold to some Pennsylvania gentlemen, the writer was employed as Superintendent, and drove a still lower level on the vein, 100 feet, out of 75 of which he took, in simply doing the dead work, ore which milled as follows:

Lbs. Ore.	Ozs. Gold.	Ozs. Silver.
67	59/100	1,002 70/100
410	38/100	290 10/100
2,753	29/100	376 54/100

(Although Weston gives very specific assay reports throughout this pamphlet, the missing factor is just how rich the ore was

that the owner picked to be shipped. By taking only the richest rocks of ore, and not shipping the ones that weren't quite as rich, the assay reports could be increased significantly.)

The other lodes of this group, two of which are on the same lode as the Yellow Rose and Declaration, before described, will be developed at great depth by simply driving ahead in the Wheel of Fortune, which intersects them. The whole group comprises 60 acres, and is in heavy pine timber, with Silver Creek, a fine, never-failing stream, running through it, and a good mill site. The toll-road also passes within a quarter of a mile of the mines. *(At this point the toll road wasn't much of a road — more of a trail.)*

When it is understood that the work done in the Wheel of Fortune vein has only explored it for about 400 feet in a horizontal direction, and nowhere gained more than 200 feet in depth, the above is a remarkable showing.

The Potosi, a northerly extension of the Wheel of Fortune vein, has one tunnel on it, driven 155 feet in ore, and a shaft down 40 feet, also in ore, the pay streak varying in width from 4 inches to 2 feet. Some six tons of ore taken out in simply driving this level on the vein milled from 108 ounces to 475 ounces silver per ton, and about half an ounce of gold to the ton. The ore is much of the same character as the Wheel of Fortune ore, except that it has associated with it pink manganese spar and also arsenical iron pyrites with a bright yellow powder, the result of its decomposition. However, this Potosi ore, good though it was, had the unmistakable appearance of "top rock" surface ore, the average depth below the surface at which it was got being about 60 feet.

Within a stone's throw of the Potosi are several promising and interesting prospects. The Stonewall Jackson is a very pretty vein, about three and a half feet between walls, in the same trachyte formation, has the same quartz as the Wheel of Fortune for its ore gangue, and the character of the ore is brittle silver, copper, and iron pyrites, and a little blende and manganese spar occasionally associated with the metals. One of its peculiarities is that where the iron pyrites join the quartz gangue, native or wire silver is frequently found. An assay of this white iron alone gave returns of 85 ounces silver to the ton and half an ounce of gold. I made a test of about six inches of the ore myself, and I yielded 120 ounces of silver to the ton. Other sampled assays have returned 131 and 145 ounces to the ton. There is a shaft down 25 feet, and all in ore. The Neptune, near by, is in a vein about 12 feet between walls, and has a shaft down 32 feet, showing upwards of 3 feet of ore-galena, white iron, manganiferous quartz, and a little copper pyrites. Assays of this ore have yielded as high as 80 ounces silver to the ton. The Highland Chief is just below the Wheel of Fortune, and is a northwest-southeast lode, and has a drift on the vein 45 feet in length, showing upwards of 3 feet of galena and gray copper ore in a gangue of white quartz. Tests of this ore have given about 100 ounces of silver to the ton.

The Ruby Trust has turned out to be one of the most promising prospects on Sneffel's basin. It has had two levels run during the past winter, the work consisting of a crosscut tunnel 100 feet long, and the length of the drifts on the vein being each 193 and 83 feet. In the lower of these drifts some very beautiful ruby silver ore was encountered, the streak varying in width from 2 to 18 inches, and containing as well as the ruby, brittle silver, gray copper, and native silver, in form of wire and leaf, and yielding from 30 to 3,000 ounces of silver and $15 in gold to the ton by recent sampled assay tests. There is a connecting shaft between the two levels of 73 feet. The vein is a true fissure, and has good working porphyry for vein

The Ruby Trust Mine and Mill were built below Yankee Boy Basin. The small mill and the smokestack of the power plant can be seen in this photo. Unfortunately, it was in an avalanche-prone area and had to be rebuilt several times over the years.

Courtesy Western History Dept., Denver Public Library. X-61970 matter, clay gouge, or selvage. *(The Ruby Trust was a valuable mine and is still worked on occasion, although it is in an avalanche zone and has been destroyed several times over the years.)*

The Eldorado is one of the prettiest lodes, owing to its still being in the hands of the original discoverers, who are men of limited means, but little work has been done on it. What work has been done, however, and it is all dead work, no stoping, has been more than paid for by the ore taken out in driving the vein, and such lodes are one in a thousand, or perhaps, more correctly speaking, ten thousand.

This mine is situated close to the end of Sneffle's toll-road, its altitude being 10,500 feet above sea level. It outcrops up the mountain side, and had sites on it for at least five levels of drifts on the course of the vein, and the lower drift would gain a depth of 500 feet below the surface at the upper end of the claim. The character of the ore is ruby and brittle silver, gray copper, copper pyrites, with a small percentage of steel galena and a little red blende, the gangue of the ores being white quartz and some baryta.

The following are mill returns of some of the ore taken out in doing the dead work:

One ton sold to Crookes' works, Lake City, 676 ounces silver per ton.

One ton sold to Ocean Wave works, Lake City, 396 ounces silver per ton.

Two and a half tons sold to Mather & Geist, Pueblo, 496 ounces silver per ton.

Three-quarters of a ton sold to Windham smelter, 248 ounces of silver per ton.

One half ton sold to Windham smelter, 138 ounces silver per ton.

One and a half tons sold to Mather & Geist, Pueblo, 496 ounces of silver per ton.

Add $11 in gold to the ton in the last-named lot. There remains on the dump about 75 tons of ore, carrying from 50 to 100 ounces silver to the ton that would not pay to ship. Is not this an instructive homily on the state of our camp and its need of reduction works, when 50 to 100 ounce ore is termed "low grade" and will not pay to ship, and where, with every facility for reducing ore, viz.: abundance of coal, lime, iron fluor, spar, timber, and water power, we have to send our ores to Pueblo, 330 miles, to be treated; and until quite lately 200 miles of that distance by wagon. *(With good transportation and reduction works, ore worth as little as $30 per ton could be mined profitably.)* The development of the Eldorado consists of three levels, 260 feet in all, on the course of the vein, and a 25-foot shaft and 75-foot open cut, and all in pay ore. It is a full claim, 1,500 by 300 feet.

On the other side of the basin from the Eldorado is the well-known Yankee Boy group of mines, comprising the Yankee Boy, Dutch Boy, York State, and Genesee. The Yankee Boy proper and the Dutch Boy are adjoining claims on the same vein, and this is another of those extremely rare claims which has paid from the grass roots, but has been worked in a desultory manner by leasers and the original discoverers, who are still owners. The York State and Genesee are adjoining claims on the same vein, and parallel vein with the Yankee Boy, being only 130 feet distant. This is an undoubted true fissure vein, from four to six feet between the walls, in trachyte formation, its course being northwest-southeast, and the character of the ore, ruby and brittle silver, associated with a small percentage of fine steel galena, and somewhat of red blende, the gangue being white quartz and

The Yankee Boy Mine was owned in part by William Weston, and it produced valuable ore (but in small amounts) for many years. Its main problem was the deep snows at high altitude. Only the dump remains today in beautiful Yankee Boy Basin.

Courtesy of Western History Dept., Denver Public Library. F12720

some heavy spar. In 1877 the owners sold twenty-one tons of ore that gave average mill returns of 350 ounces silver to the ton. In 1878 nearly 200 tons were produced, and one lot of twenty-three tons averaged, it is said, 1,231 ounces silver to the ton. In 1879 the production was small, but one lot of seventeen tons averaged 325 ounces silver to the ton. I find it hard to obtain very exact statistics, as a good deal of ore was extracted and sold by leasers and no account kept, but as nearly as I can estimate there are some 600 feet of levels of drifts in the vein, and ore to the value of over $100,000 has been extracted. This magnificent property is several hundred feet below timberline, and can be worked winter and summer with ease. The Yankee Boy vein, in my opinion, crosses the Eldorado at an oblique angle, the Ruby Trust being near by, and from the similarity of the ores and gangue of these three mines, they probably had a common origin.

At the head of another fork of Sneffel's Creek, but in the same main basin, is the Virginius mine; and this is another of those notable lodes, which, in miners' parlance, has "paid from the grass roots." The ore is of a very different character, however, to those last described, being galena and gray copper of a very high grade. The average width of the vein is about four feet, and the pay streak varies from three to ten inches, the average yield of the ore to this date being 315 ounces of silver to the ton. Briefly, with about 1,500 feet of levels and 370 feet of shafts, this magnificent mine has sold ore to the value of $75,000, with ore remaining on the dump that would not pay to ship, estimated at between three and four hundred tons, running from 50 to 100 ounces silver to the ton. The mine is, at this writing, looking better than ever, and has a pack train at work bringing down the ore that has been taken out during the past winter.

On the same vein as the Virginius and the adjoining extension is the Monongahela, showing a width of between 4 and 6 feet, and from 1 to 2 feet of ore of the same character as the Virginius. It is developed on three levels, or a total of about 650 feet, in driving which, and without any stoping, ore to the value of $3,000 was taken out. As the most reliable proof of its quality I give a few of the returns of the ore sold:

THE VIRGINIUS MINE WAS ONE OF THE HIGHEST MINES IN COLORADO AT 12,500 FEET. IT WAS WINTER FOR EIGHT MONTHS OF THE YEAR AT THIS ALTITUDE, DURING WHICH TIME AVALANCHES RAN CONSTANTLY AND TRANSPORTATION WAS IMPOSSIBLE. LATER THE MINE WAS WORKED THROUGH THE MUCH LOWER REVENUE TUNNEL.

Author's Collection.

16 sacks, 168 ounces silver and 28 per cent lead.
34 sacks, 236 ounces silver and 25 per cent lead.
20 sacks, 215 ounces silver and 45 per cent lead.
2,500 pounds, 169 ounces silver and 49 per cent lead.

(The Virginius, which was eventually incorporated into the Revenue Mine, was one of the great silver mines of the area. The mine still contains ore and waits for the price of silver to rebound before reopening.)

The next northerly claim to the Monogahela, on the same vein, is the Blue Grass, showing the same ore and developed by two shafts, each 150 feet, and at two 50-foot levels.

The next adjoining extension is the Raleigh, with a 50-foot drift on it, and showing similar ore.

These are all full claims, except the Blue Grass, which is 1,150 feet in length. Thus we have 5,650 feet on the vein, with a good showing in every opening, and proving the ore body of remarkable permanence.

The Terrible is another wonderfully good mine, but has suffered from mismanagement and litigation, or it would have made itself, I believe, as great a record as any I have mentioned. The Terrible crosses the Monongahela at an oblique angle, and varies in width from 4 to 10 feet. It is in the same trachyte formation, with a soft porphyry vein filling and clay gouge. The ore is galena and gray copper in a quartz gangue. Native and ruby silver has also been found in this ore, the native being in wires and apparently plastered onto the galena, where it joined the quartz. I have been unable to get accurate statistics of ore sold, but it has also "paid from the grass roots," and the development consists of about 700 feet of levels. The following mill returns of some of the ore sold will demonstrate its quality:

60,400 pounds, 404 ounces silver to the ton, 40 per cent lead.
7,172 pounds, 66 ounces silver to the ton, 30 per cent lead.
1,463 pounds, 343 ounces silver to the ton, 59 per cent lead.
5,818 pounds, 116 ounces silver to the ton, 48 per cent lead.
8,517 pounds, 123.66 ounces silver to the ton, 46 per cent lead.

And let it be remembered that of all the mines, claims, and prospects I have described, none of them have been worked below water line, they being drained and the ore extracted from them entirely by levels; in fact, there is not a pound of pumping or hoisting machinery in the Sneffels or Uncompahgre districts that I know of, except on the Mineral Farm. *(Pumping became a real problem and expense in the Red Mountain District.)*

Let it also be borne in mind that all the claims mentioned are comparatively mere prospects, most of them in the hands of the original owners, who haven't the means to open them up to any extent; that deep workings are unknown here; that the ore had to be shipped 300 miles for treatment, in addition to the cost of packing the ore on animals from the mines to Ouray, and the cost of producing, sorting, and treatment; that only about one-fifth

THIS *HARPER'S WEEKLY* DRAWING ILLUSTRATES PROSPECTORS IN THE RED MOUNTAIN AREA EXAMINING A "FIND" FROM THE SIMPLE SHAFT THAT THEY HAVE DUG. AN ASSAY WILL BE NEEDED TO DETERMINE THE EXACT WORTH OF THE ORE THAT THEY HAVE FOUND.

Author's Collection.

of all the ore extracted would stand this enormous expense and leave a margin of profit, the rest now lying in the dump; that the only custom mills put up in Ouray failed because the owners didn't know how to run them *(custom mills worked in connection with a particular mine.)*; and let some idiot ask us how it is we haven't produced more bullion. *(Many of these dumps were later worked at a profit when costs had gone down.)*

Crossing over the crest of the divide above the Virginius workings, by what is known as the Virginius Pass, at an altitude of about 12,700 feet above sea level, and descending the other side, you are in Marshall Basin and the San Miguel Mining District; and here is located the celebrated vein on which three adjoining claims, the Mendota, Sheridan, and Smuggler, have attained a most just notoriety. I do not hesitate to say that I believe this is, and will prove, for permanence and richness of the ore body, the most extraordinary vertical fissure gold and silver bearing lode yet discovered in the State of Colorado, and the mill-runs of ore sold, and taken out entirely in doing the dead-work, will, I think, back me in my opinion.

Here let me state what I understand by the term "dead work." In creating reserves of ore or blocking out stoping ground, tunnels have to be driven, and shafts or winzes sunk or raised first; this is what I call "dead-work," and the mine whose dead-work will pay the expense of doing the same, is, I think, about one in a thousand. The above vein, which is generally known as the Smuggler lode, is an undoubted true fissure, from 6 to 12 feet between the walls, and outcrops boldly on the surface for a distance of two and a half miles. It trends south 18 degrees east, and dips gently toward the west. Beginning at the Mendota, which is just over the divide from the Virginus, the claims on the vein going in a southerly direction are as follows: Mendota, Sheridan, Smuggler, Union, Cleveland, Bullion, and Hidden Treasure, which brings you down to the valley or bed of Marshall creek, all these being on the north side of the creek. *(These claims were eventually consolidated into The Smuggler-Union, one of the great mines of the San Juans, and eventually produced many millions in ore, being worked for over 100 years.)*

THIS PHOTOGRAPH OF MARSHALL BASIN WAS TAKEN LOOKING TO THE NORTH, AND IT SHOWS THE SMUGGLER, SHERIDAN, AND MENDOTA MINES FOLLOWING THE HUGE, RICH VEIN UP THE MOUNTAIN. THE ORE RAN THOUSANDS OF FEET INTO THE MOUNTAIN AND WENT ON FOR OVER TWO MILES.

Courtesy Western History Dept., Denver Public Library X-62254.

On the south side, the Ajax is the first, and I believe there are three other extensions or claims, the names of which I do not know. The Mendota is a full claim, 1,500 feet by 300 feet, the width between walls averaging about 8 feet, and showing as pay streak of an average width of about 2 feet. This mine is still in the hands of the original owners, and but little developed. The work done on it consists of a shaft, down 100 feet, a level from the center of the shaft, driven 85 feet, and another from the bottom of the shaft, driven 23 feet. The only stoping done is a place about 20 feet long and 20 feet high. Of the 2 feet of ore taken from this dead-work and small stope, only about 6 inches was shipped, and the reader must recollect that the ore from these mines was sent by pack animals to Ouray, at a cost of about $20 per ton (12 miles), and from there to Pueblo to be treated, 330 miles, 200 of it by wagon; and it will be easily understood that none but excessively high grade ore would stand such enormous expense. The other 18 inches of ore that was not shipped remains on the dump, and will yield from 50 ounces silver to the ton, up to the lowest mill return here given — 108 ounces. In fact, the ore has yielded at the rate of $50 to $60 per foot, the cost per foot of sinking and driving being about $16.

Mill returns from Mendota ore:

Pounds.	Ounces Silver.	Ounces Gold.	per ton.
760	473	1	"
4,098	157	-	"
1,859	113	1	"
1,107	184	1 ½	"
2,972	346	3 ½	"
783	247	1 ¼	"
2,434	420	2 ½	"

11,136	252	1 ¼	“
8,348	248	1 ¼	“
12,960	110	1 ¼	”
14,152	108	1	“
10,159	270	2	“
7,147	427	2 ½	“
12,083	130	½	“
16,199	192	1	“
2,816	305	2	“
10,777	128	1	“
9,627	125	¾	“

Being men of limited means, the owners could not do more development, especially as they were occasionally bothered with water in the mine, which it will require small pumping machinery to keep clear.

The Sheridan, the next adjoining extension, is patented, and the title is in the name of a wholesale clothing firm of Milwaukee. The development consists of two drifts on the vein — one 140 feet long, and 60 feet above that, another one 50 feet long. The vein averages in width about the same as the Mendota, and the average width of the pay streak is about 2 feet.

The superintendent now in charge states that the ore taken out in driving the above levels aggregates 75 tons, yielding over 200 ounces silver to the ton, the lower grade ore being left on the dump for the same reasons as stated in the case of the Mendota. He also states that the whole two feet of ore will average 50 to 70 ounces of silver, and half an ounce of gold to the ton; and, from what I know of the mine, I consider this a very conservative statement. The Sheridan is well situated for development by drifts on the vein, as a drift northward along the course of the vein from the lower end lines would gain a

Most of the ore from the Smuggler, Sheridan, and Mendota claims was taken out at lower elevations like the "Bullion Tunnel." A small settlement eventually grew up around the Smuggler Union workings.

Courtesy of Western History Dept., Denver Public Library X-62314

depth at the upper end lines of 150 feet. There is a narrow streak of ore in the Sheridan carrying free gold, and from this some wonderful specimens have been taken. The mine is, at this writing, being worked under lease, and turning out fine ore.

The Smuggler, the adjoining southerly claim, is 1,056 feet in length. The title is a United States patent. *(Mining claims are originally filed as "unpatented," which meant the claim could be lost if not worked. When "patented," the land was purchased from the United States and no longer needed to be worked constantly to retain title.)*

The following are mill returns of ore sold. No stoping has been done, and this ore has been taken out in running a prospecting drift, or tunnel, and raising two air shafts or winzes from this tunnel to the surface, as follows: The work done consists of a cross-cut tunnel 200 feet in length, intersecting the vein at a depth of about 100 feet below the surface, all in ore; a drift north from the end of the cross-cut, on the course of the vein, 310 feet, all in ore; eighty feet from the end of the north drift, a winze to the surface, 115 feet, all in ore. Size of drifts, 5x7 feet; cross-cut, 5x7 feet; shafts, 5x9 feet.

Mill returns of Smuggler Ore:
(Note the large amounts of gold)

Pounds.	Ounces Silver.	Ounces Gold.	
2,090	561	16	per ton
2,020	393	8 55/100	"
2,000	320	8 8/100	"
9,081	350	6 45/100	"
7,454	340 95/100	7 70/100	"
7,023	311	6 67/100	"
9,080	353 27/100	6 50/100	"
4,334	200 56/100	3 8/100	"
5,373	314 77/100	6 2/100	"
8,682	224 14/100	3 50/100	"
3,273	306 6/100	8 94/100	"
4,210	199 86/100	3 3/100	"
1,323	417 22/100	7 97/100	"
7,528	256 9/10	5 20/100	"
5,827	208 25/100	4 8/100	"
7,215	126 23/100	2 1/10	"
6,280	276 64/100	6 23/100	"
5,434	342 9/100	8 26/100	"
6,247	203	4 66/100	"
6,470	198 25/100	3 59/100	"
6,532	346 51/100	7	"
8,186	198 34/100	3 40/100	"

8,477	157 30/100	2 64/100	"
8,243	131 6/10	2 5/10	"
6,626	307 53/100	6 2/10	"
2,070	135 8/100	3 23/100	"
5,327	250 82/100	4 66/100	"

In the Smuggler workings there was from 2 to 3 feet of ore all the time, but the width of the ore streak that was sold, and of which the foregoing are all the mill returns, is an average of 6 inches. The remainder of the ore is on the dump, ready for shipment as soon as we shall have cheaper freights. The amount of this ore on the dump is in the neighborhood of 400 tons, and it will yield from 50 to 125 ounces silver to the ton, and some gold.

The character of the ore is as follows: Silver occurs native, and in large quantities both as wire and massive; also the ores of silver, known as brittle silver, silver-glance and ruby silver, associated with iron pyrites, copper pyrites, and a small percentage of galena and zinc-blende. The matrix or gangue of the ore is quartz, and what is known there as "rose quartz," a silicate of manganese. The country or wall-rock is trachyte, and the intermediate vein stuff, porphyry.

Considering the very large proportion of gold in the Smuggler ore, it is perhaps somewhat strange that it has not been found in the ore free; and it will be remarked as to how very regular is the high or low yield of gold in proportion to the high or low yield of silver. I would again call attention to the fact that all the ore sold from the Smuggler was taken out in driving 5x7 drifts and 5x9 winzes, and the ore between the drifts and the surface remains to be stoped. The average height of this back is 100 feet, length 450 feet, width of ore body say 3 feet. As soon as a small

THE MINES IN SAVAGE AND MARSHALL BASINS WERE FAR UP ABOVE THE TOWN OF TELLURIDE. AT FIRST, SUPPLIES AND ORE HAD TO BE BROUGHT UP AND DOWN BY BURROS AND MULES OVER THIS ROUGH TRAIL, BUT EVENTUALLY A FAIRLY DECENT WAGON ROAD WAS BUILT.

Author's Collection.

pumping and hoisting engine is erected at this point, a main shaft can be sunk on this chimney of ore, and more development done; as it now stands, however, it will pay from the first shot. The owners have worked the mine at this point, winter and summer, without difficulty. I have made the following rough estimate of ore in sight: The 75 tons sold averaged $508 per ton, and 400 tons on the dump, say $50 per ton, $58,000. Average yield of this 475 tons, $113 per ton. Backs 100x300 by 3 feet, 9,000 cubic feet or 9,000 tons, $1,017,000 gross in sight. Now, allowing $250,000 for producing and milling this ore, and $250,000 for possible over- estimate of size of ore-body and such contingencies as "lean" places, and we still have over half a million dollars net in sight. *(The Smuggler was filed between the Sheridan and the Union when its owner realized that those two claims were larger than allowed by law. The Smuggler ore turned out to be the most valuable of the three claims. When silver prices plummeted in 1893, the Smuggler-Union's gold values kept it working at full steam. It eventually produced many millions of dollars in ore.)*

The adjoining southerly extension of the Smuggler is the Union. About 100 feet from Smuggler end lines, the owners of the Union sunk a shaft on the vein, and at the depth of 100 feet encountered a large body of the same high-grade ore as the Smuggler. They are also running a cross-cut to intersect the Union at a depth of between 200 and 300 feet below the surface, the length of the cross-cut being, I believe, about 400 feet.

The next adjoining extension is the Cleveland, and from the outcrop of this claim I once took two samples, which assayed as follows:

> No. 1, 20 22/100 ounces of gold, 381 11-100 ounces of silver-$887 per ton.
> No. 2, 5 54/100 ounces of gold, 23 13-100 ounces of silver-$139.95 per ton.

Thus for over 6,000 feet this magnificent vein shows the same high-grade ore.

The Marshall Basin mines were large and profitable, and huge mills were built near Telluride to process their ore and send it out on the Rio Grande Southern Railroad. The mines themselves were high in the mountains at the top center of this photograph.

Author's Collection.

Some 200 of 300 feet below the Cleveland the vein is crossed at right angles by the Memphis, an easterly extension of the Pandora and Oriental. These, along with the Sneffles claim, are on a large well-defined, gold bearing vein, whose width between walls is about 12 feet. The Oriental and Pandora are now being worked by a French company, with what results I am unable to say. Not far from this point of crossing the Memphis, there runs into the Smuggler vein a lode hardly less noted than the Smuggler, and known as the

CIMARRON.

This claim is 1,500 feet in length by 300 feet in width, adjoins the Hidden Treasure and Bullion, and runs into the Cleveland. Title, United States patent. It trends south 25 degrees east, and is a remarkably strong, well-defined, and rich vein. It carries free gold in leaf and wire, native silver in both leaf and wire, and also a streak of galena and gray copper, very rich in silver. When I visited it two years ago, there were about 300 feet of drift and shaft work done on it, and all in ore of a high grade, but no stoping had been done. Some of the ore extracted in doing the dead work was sold, and I give the mill returns as the best index of the high quality ore. As in other cases, of course, the low-grade ore remains on the dump.

Mill returns of Cimarron ore:

Pounds.	Ounces Silver.	Ounces gold.	
10,653	149 19/100	14/100	per ton
739	68 21/100	3 3/10	"
2,437	112 28/100	38/100	"
210	20 22/100	23 52/100	"
2,753	97	9/100 31% lead	"
2,488	100 19/100	14/100	"

1,684	82 6/100	3 1/10	"
4,318	249 36/100	29/100	"
1,924	263 8/100	58/100	"
3,431	107 47/100	1 2/100	"
3,546	309 17/100	——- lead 22%	"
6,916	153 51/100	1 65/100	"
2,034	440 8/100	1 36/100	"
1,952	541 31/100	1 1/10	"
9,790	190 25/100	1 1/10	"

To sum up, the Mendota, Sheridan, Smuggler and Cimarron are four of those very rare mines, which have more than paid for their own development, and that in a hitherto comparatively inaccessible district. *(The Rio Grande Southern Railroad eventually reached Telluride in 1893 and took out low grade Smuggler-Union ore.)* If it was not so they would not have been worked, as the owners — Sheridan excepted — are the original discoverers, and not men of capital. The southerly extensions of the Cimarron are the Bradley, Pioneer, Columbia, Erin's Hope, and Pay Rock.

In the fall of 1880, and about six weeks before ice set in, I started out from my camp in Imogene Basin to look at some placers I had heard some very stiff yarns about, these placers being situated on the Savage Fork of the San Miguel River. Crossing the divide between my camp and San Miguel, a distance of about four miles, I came to a log cabin in a little level park of pine trees, where lived the four lucky owners of the richest placer claim ever discovered in this section. The owners were extremely polite, and showed me, I may say, a cabin full of gold specimens — small ones on the shelves and big ones under tables and bunks, one so big that it had to be rolled over, instead of lifted, to show me; not solid gold, however, but a quartz boulder full of fine gold, and in vughs or pockets, containing matted wire gold and leaf gold, the fine gold being just as thick in the rock as the iron pyrites in the porphyry vein stuff of our silver lodes. A few yards from the cabin they were picking and digging out this "float" — for that is all that it is — and shoveling it into their sluice-box, and then, when the water had washed the quartz clean, picking out the chunks that contained visible gold, and saving them for subsequent crushing in a hand mortar, the gold being afterwards panned out. There is a good deal of galena in this gold quartz, so they were not using any quicksilver. The question then arises: Where did this float come from? Going up a few feet above where they were washing, I could see from the bed-rock the striation marks of glacial action, bearing down from the direction where there is a great quartz lode called Cincinnati, not over 100 yards from where this wonderful float was found. I noticed the matter then and since very carefully, and I am of the opinion that most of this gold rock has been broken off the outcrop of the Cincinnati, and I think so for several reasons. The quartz of the placers is almost identically the same with the Cincinnati, and I compared them closely. Then going down a 20-foot shaft at the Cincinnati, I broke some of the rock and found some very fine free gold in it. Added to this, the very sharp edges and jagged points of the float quartz showed that it could not have traveled any distance.

The Cincinnati is a splendid strong vein, about 20 feet between the walls, and located in a little park in pine timber, where it can be worked comfortably the year round; but being in a flat, it will be necessary to sink on it, and of course machinery will be required. But as the present owners are the original discoverers, and poor men, they are unable to develop it, and this summer they are simply going to wash their surface ground for the float gold which, at this writing, is being still taken out in astonishing quantities in the same flat close by, by the four men above spoken of, who of course were unable to wash during the past winter on account of the ice and snow. I believe the Cincinnati will, when capital is brought to bear on it, turn out one of the bonanzas of the San Miguel Mining District.

I am now informed by a miner from there that the four owners of the placer took out last autumn upwards of $20,000 in five weeks.

I speak of the Cincinnati because I saw it myself, but there are other claims on the same vein, notably the N. W. H. Jr. and Argentine, both of which contain magnificent free gold ore in large bodies.

Returning again to Ouray, I will describe two or three of the principal mines at Bear Creek, in the Uncompahgre District and immediately contiguous to town. *(Actually they were two or three difficult miles from town, as shown in the accompanying photo.)*

Among the foremost for amount of work done is the Union Mine, about four miles from Ouray — half of that distance is a good wagon road, the remainder by trail. The altitude of the Union is 9,500 feet above sea level, and it has trachyte for its walls, averages about 5 feet in width, and trends northwest southeast. The pay-streak varies in width from 3 to 30 inches. The character of the ore is brittle, ruby, and native silver, associated with iron pyrites in a quartz gangue. Seventy-five tons of ore have been sold that averaged 127 ounces silver and one ounce of

THE TRAIL UP BEAR CREEK WAS STEEP AND PRECIPITOUS — ONE OF THE HARDEST AND MOST RUGGED IN THE SAN JUANS. IN SOME AREAS THE CLIFFS WERE 300 TO 400 FEET STRAIGHT DOWN FROM THE TRAIL. THERE WAS ABSOLUTELY NO WAY THAT A ROAD COULD BE BUILT TO THE MINES, SO ALL TRAFFIC WAS BY HORSE, MULES, OR BURRO.

Author's Collection.

gold to the ton. Upwards of 200 tons of shipping ore are now on the dump awaiting the nearer approach of the railway and cheaper freights. It is owned by a New York mining company. The development consists of a shaft 75 feet deep, a drift on the vein connecting with the shaft 225 feet, and a cross-cut 150 feet. The Union is crossed at nearly right angles by the Monitor, owned by the same company. It is developed by two levels, each 225 feet, and 100 feet in length. This vein is about 4 feet between the walls, but similar to the Union, the ore being of a somewhat higher grade, both in gold and silver. The Sivyer has a remarkable record for prospect, most of the ore sold having been taken out a shaft 60 feet deep. Some of the millruns are as follows:

	Ounces Silver	
2,300 pounds	191	per ton
2,500 "	116	"
2,125 "	122	"
17 sacks	131	"
56 "	113 ½	"
2,945 pounds	113	"
3,684 "	83	"
2,060 "	210	"
1,610 "	439 and 3 ounces gold	"
3,549 "	122 and 1 ounce gold	"

The ore is principally iron and copper pyrites and brittle silver, associated in places with galena.

The gangue is a light white quartz, and easily crushed, and it is considered here good concentrating ore.

The Silver Point is about 5 feet between walls, and is quartzite formation, and is worked at an altitude of about 10,000 feet above sea-level, but is well situated for economical extraction of its ores, being in good pine timber, and a depth can be gained on the vein of upwards of a thousand feet by drifting alone. The character of the ore is galena and copper and iron pyrites, in a quartz gangue, and carrying from 85 to 90 ounces silver to the ton, the average width of the ore being 30 inches.

The following are the yields in silver of several lots sent to reduction works: 140 ounces per ton, 104 ounces, 63 ounces, 54 ounces, 157 ounces, 158 ounces, 72 ounces, 98 ounces, 118 ounces.

The development consists of two drifts, each 135 feet in length, and a shaft down 48 feet.

A very important group of mines in this belt is that known as the Silver Link. It comprises seven full claims — Climax, Silver Link, Great Epoch (these being adjoining claims on the same vein and all patented), Rob Roy, Ivanhoe, Cader Idris, and Bryn Mawr — the last named being cross lodes to the Silver Link, and varying in width from 2 to 12 feet between the walls. The Silver Link vein is a wonderfully strong one, and shows on its outcrop a width of 25 feet from wall to wall. A mill-run of Silver Link ore gave returns of 169 ounces of silver to the ton and 26 per cent copper, and assays of the ore at the Denver mint yielded 454 to 686 ounces silver to the ton.

The development consists of a 50-foot tunnel on Climax, and a 50-foot open cut on Epoch. A cross cut tunnel is now being driven, which will cut the Silver Link at a depth of 1,500 feet below the surface, the length of the tunnel being about 1,100 feet. It is now in about 500 feet. *(The Silver Link did produce good ore but was highly inaccessible. It produced over $100,000 in ore before closing in the 1890s when silver prices dropped.)*

Among the prominent groups of mines in Bear Creek, in the Uncompahgre District, is the Flagstaff group, consisting of five full claims. The character of the ore is gray copper and galena, and the main vein is fully 12 feet between walls. Mill runs of the ore have yielded as high as 209 ounces silver per ton and 25 per cent copper. The owners of this promising group are Messrs. Story, Dresser, and Proudfoot, of Ouray. The whole group will be patented during the summer.

Adjoining the Silver Link is a group known as the Crystal group, comprising the Crystal, Friend, and Challenge — adjoining claims on the same vein — Confidence, on an adjoining parallel vein, and Grand Junction, a cross-lode to the series. Very little work has

The Silver Link Mine was located in one of the steepest areas of the San Juan Mountains. Snow piled up so deep that the cabin's left smokestack rises fifteen feet above the roof of the cabin. The cabin itself was 500 feet above the river below.

Author's Collection.

The Mickey Breene Mine spread up and down the canyon of the Uncompahgre for some distance. The mine was a good one and was worked almost to present day. Some interesting structures are still visible on the site, through which passes the Engineer Jeep Road.
Author's Collection.

been done on this group, but the veins are large, well-defined true fissures, about 10 feet between walls, and carrying ore, sample tests of which gave returns of 50 to 443 ounces silver to the ton, and half an ounce to an ounce of gold per ton. The character of the ore of this group is copper and iron pyrites and brittle silver in a gangue of bluish quartz, and contains no zinc or galena.

It is peculiarly clean ore, and especially adapted to the roasting and amalgamation process. These mines are situated in the midst of good pine timber and can be worked the year round. The average width of the ore streak in this group is 18 inches, and the veins are all true fissures. Adjoining this group is one owned by Governor Pitkin and Mr. William Sherman, of Milwaukee. The veins are very large and well-defined true fissures, and carry immense bodies of medium-grade ore. But this group is as yet undeveloped, and not being possessed of any other data, the above particulars are all I can give. *(This mine became the Mickey Breene, which was by far the best mine in this area. Most development was done in the 1890s but it has been worked off and on clear up to current times.)*

Since sending my last letter, the Superintendent of the Argentine Mine, in the San Miguel District has been here and given me the following brief description of the property:

The average width of the vein is 12 feet, is in trachyte formation, and the average width of the ore body is 7 feet. The character of the ore is free gold, in white "sugary" quartz, and a streak of galena that will average about $30 to the ton. In a 4 foot breast of ore in one of the workings there is a streak of ore 30 inches in width, in a gangue of white quartz and bluish rock that will sort to run 20 ounces silver and 68 ounces gold per ton. In addition to this there is 18 inches of galena, iron and copper pyrites, and silicate of manganese. Also 6 inches of white quartz containing galena, zinc-blende, iron and copper pyrites, and free gold in large quantities. In the 16-foot shaft the vein shows 3 feet of free gold in large quantities.

In the 16-foot shaft the vein shows 3 feet of free gold ore, and 2 feet of galena ore, associated with iron and copper pyrites. A test of 400 pounds of this ore yielded about $5,000 in gold to the ton.

However, it is roughly estimated that there is in the Argentine Mine, an average width of 7 feet of ore that mills, as shown by tests from the present workings, $150 to the ton. The work so far is simply exploration of prospecting work, no stoping having been done. This property consists of two claims, Argentine No. 1 and No. 2, having a total length of 2,250 feet by 300 feet in width, and the development aggregates 316 feet.

Another important mine on the San Miguel River is known as the Mogul copper property. This lode is a well-defined vertical fissure, about 16 feet in width, in the red sandstone formation. There are four claims, known as the White Star, Bay State, New Discovery, and Hunchback, and these can be worked to a great depth by drifting on the vein alone. The character of the ore is copper glance, "horse-flesh" ore, native copper, and the blue and green carbonates of copper in a gangue of calc spar. The ore is said to resemble very closely that of the Calumet & Hecla. The average width of the ore body is 10 feet, and I am informed by the superintendent that it will average $20 in silver to the ton, and 30 per cent. of copper. The average yield of thirty-two assays lately made was $176 to the ton.

In the Bridal Veil Basin, one mile from Pandora, there are a large number of promising lodes, and to Mr. W.F. Johnson, of Pandora, I am indebted for the following description:

Bridal Veil Basin is situated at the head of Bridal Veil Park, the stream of that name falling over a perpendicular

Eventually the steep switchbacks of the harrowing Black Bear "road" allowed access to Bridal Veil Basin from Telluride. The falls are to the right of this photo and if you look closely you can see the power plant that was built at the top of the falls.

Author's Collection.

cliff of 228 feet in height, near the edge of the trachyte formation, these beautiful falls being in plain sight of the town of Panodra. The basin is approached by a very fair mountain trail leading through the basin to Silverton, and is noted for having a splendid supply of timber, grass, and water, with good well sites; and a good wagon road can be made the entire length of the basin at a reasonable expense. The veins in the basin are gold, silver, lead, copper, and iron-bearing, and are very rich where they have been opened to any depth. They are fine fissures, with well-defined walls of trachyte. Assays from these veins have run as high as $22,500 per ton, and all they need is proper development.

OPHIR

To Mr. F. J. Cramer, Deputy United States Mineral Surveyor of Ophir, I am indebted for the following history of that bright and promising little camp, which is now rapidly coming to the fore as one of our best and richest producers of precious metals:

"Ophir is situated in a small valley between Yellow and Silver mountains. Through this valley — if such it may be called — flows the Howard's Fork, a branch of the San Miguel River. From the center of the towns numerous prospects, and at least a dozen claims that have passed this stage, can be seen. It is, no doubt, owing to this central position of the town, that this district is spoken of generally as Ophir. For some time before Ophir was started, and in fact for a couple of years afterwards, this part of the country was generally spoken of as the Howard's Fork. But, of late the name of Ophir has overshadowed the older name, and is not only applied to the town but to the district in general. The name of the mining district, Iron Springs, is only used in location certificates. Ophir was first laid out, and the first log house built, in 1878. But before this time prospectors had entered into the country and located claims. Most likely the first white man who entered into this district was named Howard, for that name was found carved on one of the trees by the first prospector. No one, as far as I can learn, knows anything about this pioneer. The fact of his existence is preserved in the name given to the fork." *(Lt. George*

Howard led a group of explorers and prospectors up the Dolores and into the Ophir valley in 1861.)

Linguist was the first prospector who located a claim on the Howard's Fork. His claim was the Yellow Mountain lode, and was located in 1875. *(Although Linquist had the first claim in the valley, the town of Ophir was started on "Big Abe" Fleming's claim.)*

During 1876 the Howard's Fork was entered upon by a number of prospectors. The Santa Cruz, Nettie, and Crown Point lodes were located. The first log hut was

The small settlement of Ophir eventually grew into a good-sized town, with ore from several of the local mines being delivered by trams to mills that were within the city limits. When the Rio Grande Southern Railroad was built at the lower part of the valley, a new town was built there.

Courtesy of Western History Dept., Denver Public Library X-11517.

built during this year by Herman Nieland, who was cutting hay on the present town site of Ophir.

In 1877 quite a mob of prospectors entered into the new district. The Nevada, What Cheer, Summit, Valley View, Grand View, Sulphurets, Staadtsburgh — all claims which have developed into mines — were located during this year. Every one of the claims mentioned shipped ore to Greene's smelter at Silverton during the year from surface workings. During this year also was located the Osceola gold lode. One hundred and some odd pounds of ore from this claim brought $500 at Greene's smelter. The story goes that a prospector who first saw this ore and could have located the claim thought the ore was brass. Seventeen men wintered on the Forks that year. That winter saw an extensive fall of snow, as Mr. Milton Evans, our present County Commissioner, measured 35 feet falling during the season. As a result of this great fall of snow the season of 1878 was late in opening. A large number of prospectors were compelled to camp on Lake Fork.

During this year were located San Bernardo, Honduras, San Juan, Kendrick, Fox, and Kent lodes on Lake Fork. In Howard's Fork were located Colorado, Caribou, Montezuma, Parsons, Honduras, Spar, and Fulton. The Parsons was located by tenderfeet, one who had only a few days' grub. One had two cents in his pocket and the other ten cents. The tenderfeet of that time are tenderfeet no longer. One being "Old Man Parsons," as he is generally called, and the other L.O. Gray.

During 1878 the Gold King lode was discovered. The pocket of rich ore taken out of this claim exceeded in richness and in quantity that taken out of the Osceola.

The Gold King and Alta Mines were located in the same basin. The Gold King went through this ridge and was accessible from Ophir or Alta Basins. The Gold King basically produced only gold; but there was a lot of it, and the mine was worked for many years.

Courtesy Western History Dept., Denver Public Library X-62364.

Ten sacks of ore brought $5,000. The richness of this claim was first discovered by Jack Munn, since killed. The Munn brothers owned the Valley View, a paying claim, and in fact at one time were interested in all the valuable claims on that part of Silver Mountain where the Valley View was situated. They were good prospectors, but not men to hold on to anything; for in '70 they had nothing on the Howard's Fork.

Eighteen hundred and seventy-nine promised to be the booming year that every promising mining camp has at some time. Early in the season over 500 prospectors were in the camp. Five saloons were distributing 40-rod, and everything indicated that the boom of Ophir had arrived. But just at this time the carbonate excitement started up at Rico. That settled the boom. Prospectors wanted carbonates, and galena and gray copper was at a discount. The result was that the boom didn't stay, and Ophir had to wait. The find of this year was the Silver Belle, a silver lode, and Red Jacket and Golden Chief, gold lodes.

In 1880 and 1881 little prospecting was done. But during these years a number of claims passed into the hands of companies. The Montezuma Company, Caribou Company, Thornbuy Mining Company, Osceola Company, Summit Company, Valley View Company, Butler Company, Nevada Company, Colorado Company, and Nettie Company were formed. The prospects were worked, and winter of 1880 and the summer of 1881 did more for the solid reputation of Ophir than any previous time in its history. The claims worked showed up in so satisfactory a manner that the future prosperity of Ophir camp is assured. This work has shown that for high-grade galena and high-grade copper Ophir need take no back seat to any camp in the San Juan.

Ophir has continually, if perhaps slowly, advanced in prosperity. The future looks promising. The mines have arrived at such a state that the output can be such that those who have invested here may receive the dividends on hopes of which their investments were made.

The following is an extract from an article by Mr. Ingersoll in *Harper's Monthly Magazine* of April of this year. Having described the journey from Silverton to the Divide between the Animas and San Miguel, he says:

"From the long, slanting niche which lets the road *(a toll road built from Silverton in 1881)* down across this broken and sliding rock, and where men are always at work to throw aside the ceaselessly falling crumbs of the cliff, one gets his first view of Ophir Gulch — a valley half a dozen miles in length, without an acre of level ground in the whole of it. This end is close by Lookout Mountain; the opposite by the lofty crags of Mount Wilson. On the north, Silver Mountain cuts the sky in ragged outline, and braced against its base Yellow mountain rises straight from the creek side to an almost equal altitude. In the crevice between stand the score or so of log cabins which constitute what may be considered the liveliest camp in the whole San Juan. It is only four years since the value of this locality was made known, but now the mountains on both sides of the gulch are pitted like a pepper-box with prospecting tunnels, and there are perhaps twenty mines ready to ship ore in profitable quantities, even under the great disadvantage of the isolation. The lodes in general run northeast and southwest, but good openings have been found all the way from the brink of the creek to the shattered combing that casts its ragged shadow down the long, white slopes.

"In general, it may be said of Ophir that, as yet, there are hardly mines there, since systematic development has in no case reached a degree that will warrant the term, but there are a great many of which ore can now be shipped a hundred miles on donkey back and in wagons to the railway and thence 200 miles to Pueblo, and still yield a handsome profit. Half a dozen gold workings in particular are very rich, and several sales have been made exceeding $50,000 for a single location. Ophir intends to stay, and is working vigorously to get out ore, so that when the railway comes to Silverton — and this is promised for May, 1882 — if it never gets nearer, she can keep pace with the other camps in the amount of her shipments." [Original Editor's note — The railway is now at Silverton, and Ophir is shipping ore largely] *(Ophir did contain a large amount of ore, but it was low-grade except for the Gold King. The railroad never came into the valley, and the town eventually moved to "New Ophir" on Colorado 145.)*

OURAY ORE SALES

I give the following statement of ore purchased in Ouray by Mr. D.C. Hartwell, agent of Mathei & Gcist's Pueblo Smelting Works, not as any guide to the amount of ores shipped, as large quantities went through other channels, but as indisputable evidence of the richness of our ores. The figures speak for themselves:

1879

Mines.	Ore in Pounds.	Ounces Gold.	Ounces Silver.
Trout	22,250	½ to 1	135 to 450
Columbia	2,250	———	100

Blackbird	8,300	———	93 to 287
Wheel of Fortune	7,500	2	150 to 884
Yankee Boy	5,300	———	181 to 396
Sheridan	37,650	1 to 3 ½	129 to 508
Smuggler	20,000	6 to 6 3/8	300 to 350
Virginius	136,400	———	310 to 781
Silver King	1,500	———	258
Mendota	6,700	1 to 1 ½	113 to 473
Cimarron	4,450	23 ½	100
Eldorado	2,200	½	202 to 473

1880

Mines.	Ore in Pounds.	Ounces Gold.	Ounces Silver.
Dexter	20,000	0.58	189
Smuggler	84,600	8.24	342
Valley View	1,000	———	325
Potosi	10,700	1.70	341
Wheel of Fortune	3,200	59	1,003
Sheridan	800	2.18	253
Yankee Boy	1,000	———	633
Cimarron	6,000	3.10	82
Monogahela	2,500	46 ounces lead.	169
Terrible	60,400	45 " "	404

1881

Mines.	Ore in Pounds.	Ounces Gold.	Ounces Silver.
Yankee Boy	450	———	541.33
" "	1,205	———	137.66
Sailors Fortune	9 sacks	———	225.16
Mendota	17 sacks	7.16	184.33
"	42 sacks	3.49	346.00
"	783	1.36	247.13
"	2,434	2.64	420.00
"	11,136	1.16	252.57
"	8,348	1.37	248.47
"	12,960	1.11	110.00
"	14,152	.63	108.00
"	10,159	1.84	270.00
"	7,174	2.62	427.87
"	12,083	.87	130.00

Mines.	Ore in Pounds.	Ounces Gold.	Ounces Silver.
"	1,619	1.02	192.00
"	2,816	1.84	305.00
"	10,777	.82	128.37
"	9.672	.77	122.22
Terrible	7,172	30 per cent. lead	66.50
"	1,463	90 per cent. lead	343.00
"	5,818	48 " " "	116.60
"	8,517	46 " " "	123.66
Smuggler	5,327	4.66	250.83
"	3,440	1.45	46.38
"	9,281	2.82	176.26
"	7,400	3.40	165.76
"	4,526	5.63	309.36
"	7,317	2.8	172.00
"	4,548	6.22	314.61
"	3,800	5.44	258.21
"	5,827	3.49	176.17
Humbolt	559	.10	202.80
"	917	———	283.00
Cimarron	1,882	———	77.00
"	1,924	.58	263.08
"	6,916	1.68	153.51
"	2,034	1.06	440.08
"	3,431	1.02	107.07
"	3,546	22 per cent. lead.	309.17
"	1,934	1 10-100 gold quartz,	541.39
"	9,693	1.16	190.57
Monongahela	1,513	23 per cent. lead.	168.00
"	2,704	25 " "	236.83
"	2,116	45 " "	215.25
"	2,360	29 " "	158.00
Pony	607	———	150.49
"	1,332	———	196.00
May Flower	3,992	———	179.00
Calliope	552	———	68.70
"	2,388	———	130.00
Flag Staff	210	———	335.70

OPHIR ORE SALES FOR '82

I copy the following from *Ouray Times:*

"From a gentleman thoroughly acquainted with the mines about Ophir we receive the following report of ore shipments from the mines of that district, almost all of which has gone to Silverton. Of some of the amounts he knows positively, others are estimates which he himself says are purposely made to come under rather than over the actual amount.

Summit Mine	$45,000	Nevada	$31,000
Silver Bell	20,000	Caribou	15,000
Alta	15,000	Parsons	5,000
Crown Jewel	5,000	Garabaldi	5,000
What Cheer	2,000	San Juan	2,000
Tip Top	2,000	Magnolia	1,000
	Total		$148,000

GENERAL REMARKS

As Ouray is destined some time soon to be a great resort for invalids as well as pleasure seekers, a few remarks as to climate may not be amiss, especially as very erroneous impressions have been formed by outsiders with regard to it. I once made a journey round the world in a sailing vessel, stopping at many countries en route, and have lived in Australia and Canada and traveled all over the United States, and am therefore tolerably competent to judge correctly; and I here assert that to me the climate of Ouray, and of her mountain peaks and basins, is the most delightful I have experienced in any country. In summer, when people are sweltering with the heat in the east, and hundreds dying of sunstroke in the big cities, here in Ouray we are cool, comfortable, and healthy, sleeping under blankets every night, undisturbed by flies or mosquitoes, and experiencing no inconvenience whatever. From about July 1 to August 1 is our rainy season, not a continuous drizzle, as in some countries, but heavy showers that come down in earnest, and sunshine between them. From May to October the hillsides up to the very flats, far above timberline, are covered with flowers, and the air is often heavy with their fragrance and the balsamistic odor of the pines. About the 15th of October the first big snow comes: this generally settles, and in a great measure disappears. Then comes the season when all nature seems drowsy and languid, and ready for the winter rest — when the mountains are more beautiful than ever with the autumnal tint of the forests that clothe them, their green, brown, and golden hues, and the deep blue of the clear sky — all softened down by the Indian summer haze. About the middle of November comes the final snow, which closes the trails over the highest ranges to quadrupedal travel and compels the biped to resort to Canadian and Norwegian snow-shoes *(today's snowshoes and skis, respectively)* to cross these lofty passes.

Then woe betide the sluggard who has not got in his supplies for a winter's work, and there are many such annually "left out in the cold."

But a winter in the mountains with plenty of mining work to do is not the hardship many imagine. For four winters my partner and I lived at an altitude of 11,100 feet above the sea level and worked in our own tunnel, combining the varied accomplishments of builder, miner, blacksmith, cook, assayer, washerman, and scribe, and I must confess that I really enjoyed it, with the exception of the first winter, when we got in late and were short of

The snow piles deep in Imogene Basin. If a horse fell off the packed trail into loose snow, it could take hours to dig him out. Sometimes it proved impossible, and the animal had to be destroyed.

Author's Collection.

"grub," having, in fact, little but flour and bacon — and that rancid — to live upon. We generally laid in supplies for eight months, as at these altitudes the trails close a month earlier and remain closed two months later than in the lower districts. We had a comfortable log cabin, well mudded and chinked, and a water-tight roof, a fat 600 pound beef and sheep hanging frozen in the meat house, with plenty of hams, bacon, dried fruits, flour, tea, coffee, onions, potatoes, cabbages, and other staples for the inner man, while our mental pabulum was represented by a good stock of mineralogical, geological, and metallurgical books, and assay furnace and laboratory, our muscles, appetites, and spirits being kept in good condition by our daily work underground. Our tunnel entrance was only about ten yards from the cabin, and so we were quite independent of the most furious storms, for, being surrounded by heavy pine timber, we were safe from snow-slides. In fair weather I used to run down to Ouray — 8 miles — on snow-shoes for my mail, and this kept up our connection with the lower world. In the winter of 1877-78 in Imogene Basin, Sneffles District, I measured the snowfall there, and it fell 32 feet. When the last snow had fallen I measured the depth in the woods, where it had not drifted, and the average depth was 8 feet, so that it settles very rapidly. This was the heaviest winter's snowfall I have known here. The severest cold I experienced was one morning when the thermometer showed 16 degrees below zero, and even then I did not feel the cold as much as I have on the coast of England, when the glass stood at 30 to 32 degrees, about the freezing point. The cold of the mountains is so bracing and invigorating that I always found it most delightful.

In the mountains of this altitude the winter lasts from November 1 to May 1. In Ouray and the Uncompahgre Park, from December 1 to March 1. In March the sun's power begins to be felt, and although there are occasional snowstorms, the ground is warm and the snow will not lie. Seasons, of course, vary greatly, but what I have stated is the general rule. This past winter was extremely mild here in the valley, and the roads about Ouray were muddy or dusty all winter.

One of the greatest sources of comfort and attraction here are the hot springs and baths. The present public baths in town are owned by the original discoverers of the springs and pioneers of Ouray, Mr. and Mrs. Washington Buchanan, and they have now six bath tubs, each about 4 feet wide, 4 feet deep, and 8 feet long, each one in a comfortable little dressing-room by itself, with looking glass, boot-jack, comb, and towels, and everything the pink of neatness and cleanliness. None but those who tried it can imagine what a treat it is, after a long tramp over the winter's snow or summer's sun, to relax the muscles and soothe the nerves by a plunge in these delightful waters.

An analysis of this hot water, made by Mr. Peter Collier, chemist of the Department of Agriculture in Washington, gives the following result: There are 99.2 grains mineral matter to the gallon of water, said mineral matter being largely composed of sulfate of lime — gypsum — with small amounts of carbonate and chloride of the alkalies. There are others, however, in different parts of the town, which contain large quantities of iron, as shown by their heavy deposits of ferric oxide.

THIS PHOTO OF THE BUCHANAN BATH HOUSE IN OURAY WAS TAKEN IN 1884 AND SHOWS FOUR OF THE PRIVATE DRESSING ROOMS TO THE LEFT AND THE BUCHANAN HOUSE TO THE RIGHT. THE NEW BRICK SCHOOLHOUSE IS IN THE BACKGROUND.

Author's Collection.

I forgot to mention in my description of Ouray that we have telephonic communication with the Western Union Telegraph Company — in fact, I once sent a message from Ouray to London, England, in the evening, and got an answer next morning. And now that the great "Scenic Line" of railway is rapidly approaching its completion to Salt Lake City and the Pacific coast, passing through the Uncompahgre Valley in its course, we shall be on one of the greatest highways of commerce and pleasure travel in the world, and "the inaccessibility of the San Juan," which has for seven long years been repeatedly thrown in our teeth, will be a thing of the past.

WESTON

[End of Series of Letters written in June 1882.]

Since I closed my last series of letters to your paper, great changes have taken place here; most important mining discoveries have been made, development has brought many of the old ones into permanent pay, the Denver & Rio Grande Railway has reached Montrose, within thirty-five miles of us, and if the "Scenic Line" is built to Ouray this summer, as there seems a strong probability of its being, mining matters here will be completely revolutionized. *(Once again Weston starts off by emphasizing the importance of the railroad to Ouray and its mines.)*

Every prospector's dream was shown in this *Harper Weekly* etching accompanying an article on Red Mountain. The publicity started a major rush of "greenhorn" prospectors — men who had no experience to show what they were looking for or the hardships they would face; yet they were all sure they would "strike it rich."

Author's Collection.

My object in writing now is to describe the really important discoveries made during last summer, and also to give a sort of general review of the past season.

Of the former, the most startling are the enormous bodies of ore being uncovered in the Red Mountain District, eight miles from Ouray, and on one of the tributaries of the Uncompahgre River, which flows through this town. I returned from the Red Mountain yesterday, and will describe the mines as they appeared to me, though of course, with the small amount of work done, no definite conciliation can be arrived at, and one must naturally be allowed to theorize somewhat. Certain it is, however, that for a camp only five months old, three of these being winter months, the amount of ore uncovered beats anything yet seen in the State, Leadville not excepted, and I believe if there ever is a camp that will approach the phenomenal richness of Leadville, this Red Mountain camp is the one. *(The Red Mountain District did not approach the output of Leadville, but it was one of the richest mining districts in Colorado.)* I am quite aware that a great deal of nonsense has been talked and written about this and that new "Carbonate camp" and this and that "second Leadville," and I therefore hesitate to tread in the same ground; but there are a great many old Leadville miners there, and their faith in the extent and lay of the ore body is so strong that they went in there in December, with the snow on the ground, and are living in tents and commencing to sink shafts. This, however, so far as shown, is not a carbonate camp, the ores being principally the sulphides of lead and copper. There are, it is true, gray carbonates of lead found associated with galena, but not in quantities sufficient to give it the name of "Carbonate camp."

Leaving Ouray for these mines you follow the course of the Uncompahgre River by the Bear Creek wagon road for two miles, and then, turning to the right where Red Mountain creek empties into the Uncompahgre River, you follow the course of the former stream for two miles more in dense pine timber. *(Even though the Uncompahgre River, which is alongside the Engineer Pass road, is smaller than Red Mountain Creek, it took its name from the fact that it ran alongside the first and only road into the area from the south.)* At this point, four miles from Ouray and at an altitude of 9,300 feet above sea level and just 2,000 higher than the town of Ouray, you enter a flat valley, or park, from a quarter of a mile to one mile in width and two miles in length, covered with a dense growth of willows and other shrubs and long grass (the latter being cut for hay every autumn). There are several small lakes in this valley, and a great deal of swamp or boggy land, Red Mountain Creek and its tributaries, among which are Grey Copper Falls Creek, flowing through it. The greater portion of this park seems to be underlain with a bed of bog-iron ore of unknown thickness. On either side of the park is a high range of mountains and trachyte peaks, that on the west being the divide between the Red Mountain District and Imogene Basin in the Sneffels District, and that on the east being the divide between the Red Mountain District and the Uncompahgre District and Poughkeepsie Gulch. I give these particulars so that any one looking at Professor Hayden's, Captain Wheeler's, or Thayer's maps may easily determine the precise location.

At the upper end of the Red Mountain Park commences the chain of scarlet peaks containing some of the richest lodes in the San Juan region. I have always

WILLIAM HENRY JACKSON TOOK THIS MAJESTIC PHOTOGRAPH OF RED MOUNTAIN PARK. MOST OF THE MAJOR MINES WERE AT THE BASE OF THE MOUNTAINS IN THE DISTANCE. EVEN IN THE MIDST OF GREAT WEALTH AND FORTUNES, JACKSON TOOK THE TIME TO ENJOY THE AWESOME SCENERY.

Author's Collection.

regarded the most prominent of these red peaks (at the foot of which are located the new bonanzas) as our volcanic center, the whole system of Snaffles *(the original — and proper spelling — of Sneffels given by the first prospectors)* lodes seeming to point or converge towards it. It will now be of great interest to recur to the following remarks on these Red Mountains by Professor Hayden in his geological and geographical report of Colorado of 1876:

"On the point of petrographic character, this group presents more attractions than any of the volcanics. Many brilliant colors of the most delicate shades are shown by the mountains and ridges. They are due to admixtures of certain mineral substances. Dark colors may be said to be characteristic of the main bulk of this group, but a very prominent exception is made by what I have termed the 'red stratum.' Originally white, the presence of ferric oxygen compounds gradually changes this color to yellow, orange, red, and brown. The rock is a microcrystalline feldspathic paste of white color, containing very minute transparent crystals of sanidite and small crystals of pyrite. Throughout the district, wherever this stratum could be traced, the crystals of pyrite were contained in it as an impregnation. Decomposition of pyrite releases the sulphur and changes the iron from a bisulphate to hydrated sesqui oxide. This in varying percentages produces the colors and shades above enumerated."

In another place he says, speaking of the discovery of mines within the limits of the same group:

"As significant [as] I regard the presence of the 'Red Stratum,' I do not mean to insinuate that such localities as do not exhibit it will be found barren of ore, but I am of the opinion that the bountiful mineral impregnation with which this group is supplied indicates the former existence of such causes as led to the formation of minerals that are classed as 'ores.'"

And in another place he says:

"In previous pages the impregnation of the 'red stratum' by pyrite has been mentioned. This mineral was probably segregated during the period of the cooling of the rock. Its presence denotes nothing save the existence and ejection of a large amount of iron and sulphur at the time of eruption. Such eruption also must have prevailed as [was] favorable to the combination of this metal and metalloid. In case, therefore, if this mineral or its component parts were drawn from the same source that furnishes the trachytes, it may justly be presumed that the resources were not thereby exhausted. Inasmuch as the 'red stratum' is one of the older members of the mountain group, it may be inferred that its source lay nearer the surface of the earth than that of the succeeding flows. Assuming that the fissures, either as such, or in ramifications and other forms, reaches downward to this source, or approximately to it, we may explain the filling of the fissures with metalliferous and other matter. Infiltration in its widest sense has no doubt produced the ores as we now find them in veins. Whether such infiltration occurred by the means of mineral substances of solution or in a state of volatilization cannot always be determined. In this instance we have a case where, perhaps, both methods were employed. From the vertical distribution of minerals in the vein much can be learned, and, no doubt, future developments of the San Juan mines will furnish [much] data of the highest interest and value."

The height of the summit of this red chain, at a point immediately above the mines, is, I should judge, about 12,000 feet above tide-water, some of the highest cones or peaks exceeding this, say 500 feet. From a point, say 800 feet below the summit, to the bed of the Red Mountain Creek (which is about 800 feet, vertical, below the Yankee Girl workings) is a succession of ridges and rounded knolls covered with pine timber, and in between these knolls deep gullies, ravines, and basins, all these having been formed apparently by the action of frost, wind, and water on the easily decomposed porphyries; and in some places where the rock protrudes, it has been

Once the Red Mountain Mining District was fully developed, the slopes of the Red Mountains became covered with mines. The Guston (left) and Yankee Girl (right) are the most prominent of the mines in this photo.

Author's Collection.

changed into all sorts of fantastic forms and different colors. The rock overlying the ore beds already exposed is a soft cream colored porphyry, more or less stained with ferric oxide, this apparently merging into a quartz-porphyry as the ore is reached, this same quartz-porphyry being the only gangue I could see in the ore, together with somewhat of talc in foliated masses and as a white powder. Immediately above this soft porphyry overlaying this ore body I observed a zone of genuine trachyte which appears to me to be about 50 feet in thickness, and from this to the summit of the red chain before mentioned, it seemed to me to be all porphyry more or less of trachyte in places, or modifications of the same igneous masses. In the shaft-house of the Orphan Boy I had a fine opportunity to examine the country rock through which they had sunk and got solid ore; it was genuine porphyry, and they struck the ore in sinking twelve feet. What is underneath the ore I cannot say of my own observation, as the snow was too deep and my time too short to admit much prospecting; but the miners and prospectors there told me that it was limestone, and from what I know of the geology of the country, I am inclined to believe this is correct.

The ores are of two kinds — galena and a copper ore that looks to me like an automonial copper glance. That from the Hudson being of a very dark grey color, in fact nearly black; that from the Yankee Girl being of a grey color and associated with copper pyrites, and that from the Orphan Boy having the usual beautiful colors of erubescite, and all these having the appearance of solid masses of metal, without gangue in them, and not having the combed or banded texture noticeable in all the ores of our vertical fissure quartz veins. I send you specimens herewith commencing at the lowest point (as you ascend the valley from Ouray) at which I saw ore, the altitude being 10,000 feet above sea-level; half a mile further up are the Guston and Robinson mines, altitude 10,200; 10,350; then a mile and a half further up the gulch, the Hudson, 11,100; and 2,200 feet further up, Congress, 11,300.

So, taking everything I saw into consideration, I am of the opinion that the whole thing is an immense horizontal contact lode, having its outcrop on the side of this chain of Red Mountains, dipping slightly towards the east, or into the mountain, and pitching downward into the earth as you travel from south to north, or down the valley toward Ouray; of course, subsequent development may prove them to be irregular segregated contact ore deposits, but in the meantime I incline to the former belief. True it is that similar ore has been discovered in the Roger and Champion considerably above the line described by the altitudes given above, but I have no doubt that the flat lode, if it is such, will be found to be faulted in places by upheavals subsequent to its formation. I was unable to visit the two last named mines. *(It was eventually discovered that although there was a lot of low grade ore en masse, the really rich Red Mountain ore was in chimneys of 10 to 50 foot radius and extending vertically into the ground with the ore getting less desirable with depth. Still, the Red Mountains were one of the great mineral discoveries in Colorado.)*

In the summer of 1881 John Robinson, A. Meldrum, A.E. Lang, and A. Deitlaf discovered the Guston mine, but the ore proved at that time too low grade to ship; and so that winter they went out, returning again in the spring of 1882, and Mr. D.C. Hartwell, agent for the Pueblo Smelting and Refining Company in Ouray, being anxious to obtain lead ores, kept them at work developing the Guston. On the 14th of August last, Robinson took his rifle and went off after deer, which abound during the summer in Red Mountain Park, and while sitting on a log, picked up a boulder, and being astonished at its weight, broke it in two and found it to be solid galena. He then went to work and soon discovered the enormous ore body now known as the Yankee Girl mine. On the 20th of September last, the Yankee Girl was sold for $125,000 cash, and the four lucky owners are now enjoying a rest in their old homes "back East." *(The prospectors weren't too lucky. Their discovery eventually produced over 100 times what they sold it for.)* As there were no sides or

IN THE 1890S, A SMALL TOWN OF ABOUT 300 EVENTUALLY SURROUNDED THE GUSTON AND ROBINSON MINES IN THE RED MOUNTAIN MINING DISTRICT. THE MINES THEMSELVES ARE BEHIND THIS CLUSTER OF HOMES.

Author's Collection.

bottom to the ore, they could not determine how the ore lay, and so staked off two other claims adjacent, the Robinson and the Orphan Boy. The latter turned out, as nearly as I could ascertain, as big as the Yankee Girl, as they struck ore 12 feet below the surface, and then the shaft went down in solid ore for over 20 feet, until water stopped them.

The Yankee Girl has two shafts, 42 and 50 feet deep. No. 1 shaft is all in solid ore, showing a bed of ore over 40 feet in thickness and no bottom reached. In No. 2 shaft, which is above No. 1 on the hillside, there is over 4 feet of ore. A drift has been run connecting with the bottom of No. 1 shaft and is being driven onward to connect with No. 2. A stope is now being taken out at the bottom of No.1, just above the drift, and I went up into it and saw a

This is one of the earliest photos of the Yankee Girl Mine — taken about the time that Weston was writing. Shown in the background, getting ready to carry the day's ore production to the railroad at Silverton, is the large string of mules that Weston wrote about.

Author's Collection.

room that I should judge to be at least 10 feet square, all in solid ore, and two men only stoping out the ore, which they are shipping by mule train to Silverton at the rate of ten tons per diem. This ore is netting the owners from $55 to $60 per ton in St. Louis. It will give some idea of the ore body and its solidity when I state that there are two men stoping and two sacking, and these keep seventy-five mules making a round trip a day from the mine to the end of the Silverton wagon road, each mule carrying about 250 pounds of ore. No sorting is done, the ore being simply broken up small enough with a rock hammer, and then shoveled into sacks.

The Guston is only about 300 yards from the Yankee Girl, and where they discovered it at the mouth of the present drift the galena sticks out of the bank, at the foot of a porphyry knoll, in big boulders. The owners have driven a tunnel 40 feet in this ore, and I could see no signs of a wall, the ore being solid in sides, roof, and floor, and over 6 feet wide in the breast. About 15 inches of this is massive copper pyrites, and the rest galena, mostly solid; but where there was a little gangue mixed through it it was white. There are upwards of 100 tons of ore on the dump, which were taken out in running this 40-foot drift. This mine is still owned by Robinson and his partners, the former owners of the Yankee Girl. The thickness of the ore bed of this mine cannot, of course, be determined with only one drift and ore in every side. *(Although, as already mentioned, the Guston got a late start; it eventually became one of the great mines of the San Juans, producing over six million dollars in ore.)*

Among other promising claims in the neighborhood of the Yankee Girl group are the following:

Yankee Blue, adjoining the Yankee Girl and a parallel location; owners Buenemann, Terpening, Shaw, and Warren. They are now sinking a shaft 153 feet from the Yankee Girl shaft, and are down about 30 feet, with good indications.

Jackson and Kokomo. These are adjoining claims to the Yankee Blue in the south and Orphan Boy on the east. Same owners as Yankee Blue.

The Richmond is on the east fork of Red Mountain Creek, 1,000 feet from the Yankee Girl shaft: owners Buenemann and Terpening.

The Jay Gould and Vanderbilt are on a strong vein carrying galena and copper, and intersect the Genesee; owners Buenemann and Terpening.

The Pittsburg is also a promising claim, owned by Parsell, Buenemann, Terpening, and Lowe.

Also adjoining the Yankee Girl and Orphan Boy lodes, and extending across Red Mountain creek, are the Joaquin Miller and Gulch lodes. North of the Gulch lode lies the J. B. Grant lode, which is a large vein showing on the outcrop mineralized matter for over 40 feet. West of the Gulch lode is the Ranchman. The above four lodes are owned by H.G. Carson, M.W. Dresser, D.F. Day, J.R. Letcher, R.W. Haskins, Major Call, and W.E. Block.

The Robinson, close by the Guston, has also a short drift run into the base of a porphyry knoll, and they took out some galena and gray carbonates from the roof of the floor, but the owners seem to have got such a big thing in the Yankee Girl that they are apparently disposed to leave their other two claims, the Orphan Boy and Robinson, alone till summer.

In the Little Maud, owned by McDaniels, Ballou, and Morgan, of Silverton, just above the Orphan Boy, they had sunk a shaft, I think, 30 feet, and water stopped them just as ore began to come in at the bottom.

The Hudson shaft, down 52 feet, has exposed a body of ore 11 feet in thickness, and I was credibly informed that the mill runs average $90 to the ton in gold, silver, and copper; and there were, I should judge, upwards of 200 tons on the dump. This shaft, like many others, was commenced on top of a knoll, and solid ore was struck twelve feet from the surface.

The Congress I did not see, as I was told the shaft was full of water; but the *La Plata Miner* states that the ore body is 18 feet in width, and 110 tons of the ore was shipped which netted $100 per ton in Silverton —

The Congress Mine, which was located on the San Juan County side of the divide at Red Mountain, was ultimately developed into a significant operation surrounded by a small town, including the nice residence shown in this photo.

Author's Collection.

copper, silver, and gold. It is being opened by a cross-cut tunnel.

All these mines are in Ouray County and on the Ouray side of the Divide, between the waters of Red Mountain Creek and Mineral Creek. *(Actually, some of the mines like the Congress were in San Juan County, but the vast majority of successful mines were in Ouray County.)* The distance from Ouray to the Yankee Girl mine, as accurately as I can tell by walking it, is eight miles at the outside; from the Yankee Girl mine to the summit of the Divide it is at least two miles, and from the Yankee Girl mine to Silverton at least twelve miles. Captain Wheeler's United States Geographical Survey Atlas fixes the distance from Gray Copper Falls Creek to Ouray as six miles, and the Yankee Girl is certainly not more than two miles from that creek.

A great deal of unnecessary wrangling is being indulged in by the local papers as to whether these mines are tributary to Ouray or Silverton. Nature has decided that point by placing the mines on the Ouray side of the Divide, the Yankee Girl group being at least 1,000 feet lower than the summit of the Divide, as the altitude by my aneroid barometer at the Yankee Girl is 11,300 feet *(The pass is 11,075 feet where the road crosses today. Weston was probably using the elevation of the pass at Red Mountain City, about a half mile east of the present road.)*. At present, the ore is going to Silverton because Silverton has the railway and Ouray has not, and the ore will keep on going to Silverton until the Denver & Rio Grande Railway builds to Ouray, when every pound of ore will come here and all the miners' supplies will be purchased here. Every one knows that the natural way for ores to find a market is down hill, especially when the distance is shorter; and if the Denver & Rio Grande builds to Ouray, unless that corporation should discriminate in rates against the Ouray branch in favor of Silverton, supplies can certainly be laid down here, the way they now come, cheaper than at Silverton, as the distance to the Red Mountain mines via Gunnison and Ouray is 137 miles shorter than via Alamosa and Durango and Silverton. More than that, the main line from Denver to Utah will be completed in a month or so, and we shall then have flour, fruits, and other supplies from Utah and the Pacific coast direct. If any proof of this difference in length of route is needed, I refer to the Denver & Rio Grande Railroad maps.

Ouray is the shortest and cheapest route to the mines. It seems only natural that if the output of this new mining district should be as enormous as is confidently expected, the Denver & Rio Grande Railway will prefer to build a branch to Ouray, 32 miles from Montrose, on the main line to Utah, rather than carry the ore out by a roundabout route 137 miles out of the way, especially through a wretched country where there is comparatively no local traffic. The saving of a dead haul of 137 miles means considerable economy is the matter of wear of rolling stock, road bed and rails, fuel, trainman's wages, etc.

I am told that Silverton has entered a protest against Ouray's petition for the railroad. If this be true, we need no better proof that this is the natural channel for the ore, and that Silverton knows it. *(Ouray finally got the extension of the D&RG Railroad in 1887. For one year the railroads were about equal distance to Red Mountain from Ouray and*

Silverton. However, in 1887 Otto Mears began construction of the Silverton Railroad from that town to Red Mountain. After that time almost all ore went out through Silverton.)

One of the best claims in the above district, in addition to those already mentioned, is the Humbolt, owned by Sheriff Parsell, John Jardine, cashier of the M. M. Bank *(Miner's and Merchant's Bank)*, and Jasper Brown, all of Ouray.

A drift was started at the base of a knoll about a mile north of the Yankee Girl, and driven 15 feet, all in ore. They then went down 60 feet lower, and commenced another tunnel to run under the first. This latter they have driven 30 feet, and is also in ore, but not so solid, perhaps, as the upper, which has ore on roof, floor, and sides. In the upper drift the ore assays from 39 to 49 ounces silver and 23 per cent copper to the ton, and in the lower drift 51 ounces silver to the ton and 31 per cent copper.

The over lying rock is porphyry, and the gangue quartz-porphyry, and the ore, I think, is tenorite of the black oxide of copper. The altitude of the Humbolt is 10,000 feet above the sea level, and on the direct line of all the other ore bodies.

The Genesee, a magnificent claim, is also owned by Sheriff Parcell and Jasper Brown, of Ouray. It shows a solid body of galena three feet in thickness and associated with copper, and it yields from 15 to 40 ounces of silver to the ton. It is about 1,000 feet southeast of the Yankee Girl, and its course appears southwest and northeast. This, however, has somewhat the appearance of a vertical fissure. The work done consists of a 20-foot shaft and a 15-foot shaft, all in solid ore.

A little further up the main Red Mountain gulch are the Gertie H., Raven, Erie, and Maud S., all owned by Al

Dutton and Herman Robe, of Ouray. The first two show very handsome galena and copper ore. Adjoining the Yankee Girl is the Pennsylvania — same owners — and another adjacent claim to the Yankee Girl is the J.R. Dutton, owned by Dutton, Burt, and Penover, of Ouray. In fact, all the ground immediately adjoining the Yankee Girl group has been taken up, and is being worked. The Little Annie and Smuggler, owned by Joe Wallace, Alling & Co., and Wm. Schultz, of Ouray, are adjoining claims to the Guston and Robinson, and the White Cloud, half a mile north of the Guston, is being worked by the same owners; and some very pretty copper ore has been taken out in sinking a shaft just commenced. The Morning Star is owned by Wallace and Schultz, and the Red Mountain claim, owned by Wallace, Hoover, and Schultz, butts up against the Guston, and is a good one. The American Girl, owned by Baker and Williams, of Ouray, is half a mile north of the Yankee Girl, shows a pretty lead and copper float, and a short drift run on this claim shows the soft white porphyry usually overlaying the ore. The Jackstaff, adjoining the American Girl, is a full claim, and owned by Price and Richardson, of Ouray. A cabin is being built by the owners, who will do systematic prospecting work on their claim during the winter. Randall and Spence and Murphy, of Ouray, have adjoining claims to the Jackstaff, and it is considered first-class ground.

An adjoining claim to the Orphan Boy is the Union Trust, owned by Penoyer, Burt & Burnett, of Ouray. They have sunk two shafts of 23 feet each, got ore, and are now running a cross-cut tunnel which is in 60 feet. They also own the Gould and Curry, with a 30-foot drift; all good claims in the center on the Bonanza group.

About a mile beyond the Yankee Girl is a claim that I believe will turn to be out one of the largest and richest. *(The Yankee Girl turned out to be the premier mine of the Red Mountain District. The Laura Dunmore didn't even come close.)* It is down in the heavy timber, near the main gulch, and is called the Laura Dunmore. There is a 4-foot breast of ore in it, samples of which have yielded 100 ounces of silver to the ton, and about 25 per cent copper. It is owned by Richard Cogar and L.E. Church, of Ouray, and will be worked directly when snow ceases to fall.

There are a hundred other claims taken, but it will not interest your reader to give the names.

Among the promising claims owned by Ouray men, I forgot to mention the Highland Chief, owned by Reed Brothers, civil and mining engineers. They are doing systematic exploration work, and have commenced a shaft, which is already down over 20 feet. The Highland Chief adjoins the Orphan Boy.

Opposite the Yankee Girl group, and on the west side of Red Mountain valley, nearly all the main true fissure veins of Imogene Basin and Sneffles Basin in the Sneffles district come through, converging or pointing, as I have said before, to the principal Red Peak, which is just south of Gray Copper Falls Gulch. In Commodore Gulch, which comes down from the divide between Imogene Basin and the San Miguel into Red Mountain valley directly opposite the Yankee Girl, there has been discovered a big vein on which are located two claims, the Iranistan and the Bobtail. On the Bobtail, which is the upper one, a shaft has been sunk 12 feet, showing a large body of galena; above this an open cut on the vein shows four feet of galena, copper, and iron; 400 feet above that again another working shows six feet of galena, copper, and iron ore; and 400 feet from the upper end there is a body of galena 10 feet in width, the vein at this point being 20 feet in width. The Iranistan has a 25-foot cut 150 feet from the west line showing four feet of galena, copper, and decomposed matter, and in a 12-foot open cut, 100 feet lower down, there is a large body of copper, galena, and decomposed matter. There is good timber on both claims. Running parallel with the above, other vertical fissures have been discovered. One of these, the Comet, is 7 feet between walls and shows 3 feet of galena and quartz. The Great West is an extension of the same, and in the next gulch north is the McIntyre, about 2 ½ feet between walls, these being owned by W. H. Fay, Rory McIntyre and Arthur McIntyre, of Ouray.

Since commencing this letter, I have received information that no less than four towns have been laid out in the Red Mountain district, and the excitement is intense. *(There were actually seven towns, but the main ones were Red Mountain Town, Red Mountain City, Congress, and Ironton.)*

Since I closed my series of letters to you in June last, the Denver & Rio Grande Railway has been completed to Montrose, and Barlow & Sanderson's four-horse coaches leave there in the morning between 7 and 8 and arrive at Ouray between 2 and 3 o'clock the same day. They leave Ouray for Montrose at noon and arrive between 6 and 7 p. m. We have also Dave Wood's fast freight line, and his six-mule teams come and go out daily.

The following are statements of ore bought by Mr. D.C. Hartwell, agent for the Pueblo Smelting and Refining Company, since my June letters were sent to you:

From the Virginius mine he bought 175 tons of ore, which averaged about 250 ounces of silver to the ton and a large percentage of lead. The following are the two highest and lowest mill certificates:

Pounds of Ore	Per cent. Lead.	Ounces Silver per ton.
4,571	40	863
10,928	43	176

Of course, as I have before stated with regard to many others of our mines, there are several hundred tons of a lower grade ore than the above lying on the dump of the Virginius awaiting cheaper freights.

WHEEL OF FORTUNE

Since July, 1882.

Pounds Ore.	Ozs. Gold per ton	Ozs. Silver per ton	Pounds Ore.	Ozs. Gold per ton	Ozs. Silver per ton
1,825	.58	277.65	17,473	.29	109.37
1,663	.43	214.54	30,572	.24	152.60
1,022	.24	122.14	16,999	.53	169.20
4,729	.14	169.20	23,879	.58	165.08
18,881	.29	112.98	24,000	.25	172.00
2,881	.58	275.9			

SMUGGLER

Pounds Ore.	Ozs. Gold per ton	Ozs. Silver per ton	Pounds Ore.	Ozs. Gold per ton	Ozs. Silver per ton
6,485	2.04	143.78	6,600	7.60	331.90
2,573	4.18	211.65	5,800	3.59	181.90
6,126	3.20	181.70			

YANKEE GIRL

Pounds Ore.	Per cent. Lead.	Ozs. Silver per Ton.	Pounds Ore.	Per cent. Lead.	Ozs. Silver Per Ton.
4,554	.56	88.66	20,233	.54	90.40

THE SNOW COULD REALLY PILE UP IN MARSHALL BASIN IN THE WINTER, BUT AS IS SHOWN IN THIS PHOTOGRAPH, WORK STILL WENT ON INSIDE AND OUTSIDE THE MINE YEAR-AROUND. THE MEN IN THE FOREGROUND ARE WORKING ON THE MENDOTA.

Courtesy Western History Dept., Denver Public Library.

RUBY TRUST

Pounds Ore.	Ozs. Gold per ton.	Ozs. Silver per ton.	Pounds Ore.	Ozs. Gold per ton.	Ozs. Silver per ton.
5,879	1.16	228.67	1,794	———	111.99
7,766	.58	135.90	3,640	.23	127.00
1,724	.96	610.61			

MENDOTA

Pounds. Ore.	Ozs. Gold per ton.	Ozs. Silver per Ton.	Pounds Ore.	Ozs. Gold per ton.	Ozs. Silver per ton.
6,255	.65	217.6	6,640	.2	87.5
6,978	1.7	319.1	9,917	.1	101.1
10,868	.8	161.3	7,175	.3	116.3
10,346	.91	150.2	6,090	.1	117.2
2,177	1.6	360.5	4,615	.1	105.2
7,766	.6	164.2	8,094	.4	127.2
8,269	.5	162.2			

The above Mendota ore went to San Miguel City, and was sampled at the sampling works of Mr. W.H. Strout there.

The following from the Mendota came to Ouray, and was sampled at the Munn Brothers Sampling Works here:

Pounds Ore.	Ozs. Gold per ton.	Ozs. Silver per ton.	Pounds Ore.	Ozs. Gold per ton.	Ozs. Silver per ton.
10,614	.80	161.3	2,121	1.06	360.5
10,106	.91	159.1	14,000	.46	209

CIMMARRON

The lots of ore given below went to San Miguel. The following is the yield per ton:

Pounds Ore.	Ozs. Gold per ton.	Ozs. Silver per ton.	Per cent. Lead.	Pounds Ore.	Ozs. Gold per ton.	Ozs. Silver per ton.	Per cent. Lead.
4,097	1-25	167.1	28	3,900	1.2	219.1	22
6,231	1.1	210.5	36	3,815	1 1/12	285.2	—
5,870	1-20	197 3/20	30	3,925	1.4	320.1	—
4,487	1.1	178.5	32	2,457	.9	315.1	—
4,740	.7	198	33	5,419	4 15/20	283.	22
4,671	1.3	195 1/5	—	5,120	1.7	320.	23
4,405	17/20	112.1	28	2,396	—	538.30	33
4,101	1 5/20	98.2	25	9,174	—	264.40	30
1,348	1.9	290.1	42	852	1.9	504.1	—
829	1.7	131.1	—	1,860	1.1	162.2	—
4,042	1	287.2	36	6,614	1.4	139.4	—
5,997	.16	71.5	—	7,857	.4	479.4	—

The Cimarron Mine is being worked under lease during this winter, and so is the Sheridan. The Smuggler and Mendota and Union, all on the same vein, would also have been worked extensively this winter, but a sale to English capitalists has been on foot for some time; and while that was on the tapis work was temporarily suspended. The well-known mining engineer, Mr. Percyval Taylor, of London, England, examined this property last

A large shaft house, tramway, and mill were eventually built at the Cimarron Mine (shown here in the 1930s). The workings were slightly below the steep and rugged wagon trail to Tomboy Basin.

Courtesy Western History Dept., Denver Public Library. X-61030.

month, and we have not yet had time to know the result of his report.

SHERIDAN.

Pounds Ore.	Ozs. Gold per ton.	Ozs. Silver per ton.	Pounds Ore.	Ozs. Gold per ton.	Ozs. Silver per ton.
3,525	2.1	201	5,636	1.2	134
2,736	3.	225	2,496	1.68	169
19,620	1.4	112	6,031	1.75	171
7,440	2.	111	5,027	5.	458
5,802	1.3	103	6,528	1.75	165
2,375	1.5	98	6,054	1.25	169
8,362	1.5	128	6,173	2.2	169
15,730	1.	102	8,105	2.	174

Pounds Ore.	Ozs. Gold per ton.	Ozs. Silver per ton.	Pounds Ore.	Ozs. Gold per ton.	Ozs. Silver per ton.
9,884	1.1	120	6,035	1.86	181
5,636	1.6	145	4,980	1.8	140
2,005	4.75	314	14,918	2.5	182
21,010	1.3	106	4,363	1.8	176
8,955	2.4	173	5,012	1.8	183
7,035	.76	94			

Up the Uncompahgre some solid winter work is being done. The Silver Link cross-cut tunnel, which will tap their group of mines 1,500 feet below the surface, is being pushed steadily ahead, and is now in over 700 feet.

The Crystal group is being worked all winter with six men, the owners having disposed of a quarter interest to a gentleman in London, the purchase money going into development work. A body of ore two feet in width was exposed which, sacked up as shot out, ran $180 to the ton, and a cross-cut tunnel is now being run which will tap this ore body at a considerable depth, so as to stope and ship the ore when spring opens.

The Cincinnati mine, as I predicted, has shown up wonderful free-gold ore in place in the vein, and at this writing the owners have got it in the bottom of their shaft, which they are sinking this winter. Some extraordinary specimens were knocked off the outcrop of the vein, and I have a piece of quartz in my cabinet which weighed about 70 pounds and which is estimated to contain gold to the value of $300. I have always believed this mine to be a bonanza, and shall continue to believe so until it is proved otherwise. The owners will wash the surface ground of the lode for gold next summer, as upwards of $40,000 has been washed during the last two summers by four men from the adjoining ground.

The San Miguel Concession, Limited, a company formed in London, Eng., owns a group of six mines in Marshall Basin, with Mr. W. Maclachan, of London, Eng., as resident engineer. I have not seen this group, but I am credibly informed by a friend who has that they have a really first-class property; one of their veins that he saw is a strong, well-defined true fissure vein, carrying a large body of concentrating ore and a streak of extremely rich gold ore. As they are only doing development work as yet, I do not think they have had any ore milled, but I see they publish in their prospectus a certificate of W. W. Munn, of the Ouray Sampling Works, of a sack of their Little Mary ore, 39 pounds weight, that ran 29 36-100 gold and 51 ounces of silver per ton. It is extremely gratifying to see English

Part of the crew at the Ute-Ulay Mine posed for the photographer. Most of the men seem to be blacksmiths or carpenters working on the addition at the right. This building also served as a powerhouse.

Author's Collection.

capital looking this way, and there are a lot of us who have been mining here for over six years who intend that they shall get full value for their money. One of the largest mining concerns in San Juan is at Lake City, in Hindsdale County, adjoining this county (Ouray). The Crooke Mining and Smelting Company, owning the Ute and Ulay mines, which were purchased by English capitalists on the report of Mr. T. Currie Gregory, C. and M. E., F. G. S., of 52 Queen Victoria Street, London, E. C., an engineer of the highest standing in England; and the two mines have since the purchase, more than backed up his report. Mr. Gregory has been in Ouray, and will return here in April or May to look over this country in the interest of his London clients. *(English capitalist were responsible for a great deal of the money that flowed into the San Juans. The Crooke brothers in Lake City were some of the first to invest serious sums of money in the San Juans.)*

Mr. J. Taylor, another well-known mining engineer, of the firm of J. Taylor & Co., 86 London Wall, London E. C., examined the mines of Imogene Basin, in the Sneffels District, and also paid a visit to the Smuggler, in Marshall Basin.

Messrs. Munn Bros., of the Ouray Sampling Works, shipped about $10,000 worth of ore, $6,000 of which was ore from the Monongahela Mine, in which they are owners. The Monongahela is an extension of the Virginius, in Sneffels Districts, and carries the same class of ore. Mr. Hartwell bought and shipped about $85,000 worth of ore from Ouray, and about $150,000 worth of Ophir ore went to Silverton.

The Yankee Girl, which is an Ouray county mine, is shipping 10 tons per diem to Silverton, and I have no data of ores sold in the town of San Miguel. So it is quite safe to say that the output of Ouray county is over rather than under $350,000, the amount stated by the *Ouray Times* in a recent issue, and the average of the ore shipped from Ouray is $197.50.

To sum up: The results of our mining season of 1882 in Ouray county are not only immensely encouraging, but have, in my opinion, definitely brought us "out of the woods," and firmly established the fact of our future prosperity. To prove this assertion, I lay particular stress on two very important facts. One is that our output of ore has so far come from the merest surface workings, as in no case can I name one of our producing mines whose workings exceed a depth of 400 feet below the surface. The other, and what is of infinite importance when taken in connection with the former, is the fact that our ores have, in the most marked manner, increased in their yield of gold and silver as depth is gained. Now I am perfectly aware that it is a common thing for owners of prospects to say, in speaking of their ore or surface indications, "it will get better as you go down." But there is no general rule of this kind that will apply to the ore in vertical fissure veins. In Przibram, in Bohemia, the lodes increased in richness to a depth of 1,200 feet, and remained constant to a depth of 1,800 feet; while on the other hand in Oruto, in Bolivia, the argentiferous lodes were wonderfully rich in their outcrop, but became poorer with depth and ultimately passed into ores containing no silver whatever. But in our Ouray county mines, the universal result of depth gained has been an increase in the yield of gold and silver in the ores. I could cite very numerous instances, if necessary. One of the most remarkable is the Declaration

This view of Potosi Mountain was taken from Sydney Basin across the valley. It shows well the rugged and steep mountains of the Sneffles Mining District. The Ruby Trust Mine and Mill is at the bottom center of the photo.

Author's Collection.

Lode, in Imogene Basin, Sneffels District. The outcrop shows a big body of course iron pyrites, zinc blende, and galena, yielding but little silver, while the underground workings, 230 feet below this outcrop, show 3 feet of solid ore — bright galena associated with copper pyrites and carrying large quantities of native silver and silver glance. In the Sheridan, the deeper they go the richer the ore is. In the Virginius, the same rule obtains; and so with a hundred others. True it is that many of our lodes show extraordinarily rich ore on their outcrop, but in every instance that I have seen of this it was where the lode had been eroded or worn down by glacial or other action. In the Ruby Trust, in Sneffels District, they had nothing on the surface, but 100 feet below it they encountered ruby and brittle silver ore, some of which actually milled 610 ounces silver and nearly an ounce of gold to the ton.

In the Altoona, in Sneffels, where they have just struck it rich, they had nothing on the surface, but a 40-foot shaft exposed 18 inches of ruby and brittle silver ore.

In the Fairplay, in Sneffels District, owned by John and Frank Carney and James Hinchie, of Ouray, another important strike has been made. From an open cut in the vein they have shown a 12-foot face with upwards of 5 feet of a dark colored quartz resembling tellurium rock somewhat. A piece of this is as near the average as could be taken, as far as appearance goes, was assayed and gave returns of 492 ounces silver to the ton, which, although we do not expect the whole 5 feet will run like that, is decidedly encouraging, to say the least.

While on the subject of discoveries, I may mention three important ones made in the Upper San Miguel district last season.

The Albion, at the head of the Cornett Basin, a supposed extension of the Cimarron, carries galena and copper pyrites, assaying well in gold and silver.

The Savage is a promising prospect, located in November last, in Marshall Basin, near the Mendota.

The Emily, in Marshall Basin, a parallel vein with the Sheridan about 400 feet distant, was located in November, and is being worked all winter. It is from 5 to 7 feet between walls, with an average pay-streak of 2 feet, carrying gold, silver, and copper. The mine is being developed by tunnel, and makes a fine showing, bidding fair to take rank at once among the paying mines of this famous basin. It is owned by D.C. Hartwell and others of Ouray.

Even our gold veins, concerning which some have got the erroneous idea that they are the richest on the surface, have proved this a fallacy, for the Argentine Mine, mentioned in previous pages, shows a richer gold ore 100 feet below the surface than on top.

Von Cotta, in his treatise on ore bodies, says: "The idea that gold veins are richer near the surface than at considerable depths must now be given up, at least as being a universal law, the lodes at Grass Valley, California, being as rich 1,400 feet below as at the surface."

There are instances on the Sneffels District where we have encountered poorer ore with depth than was to be found nearer the surface; but that is easily explained. The influence of the wall rock on the contents of the lode is well known among miners, some country rock being, as it were, inimical to the bearing of rich ore in the vein, while other country rock is conductive to it, or what Von Cotta calls a "good ore carrier." Now, when we did have poor ore with depth, I noticed that it was almost invariably when we were in a "pinch" or in a zone of trachyte conglomerate, and that when the vein widened, or we left the conglomerate and again entered the genuine trachyte, the ore regained its richness. *(Weston goes to extremes to try to prove that San Juan ore, contrary to the average, became richer with depth. Unfortunately the general rule prevailed and the ore generally got poorer. However, dreams of richer ore with depth drew investors!)*

Another evidence of an increased prosperity is the decreased cost of turning our ore into dollars. In 1877 we paid $45 per ton for treatment of ores at local reduction works, and from $20 to $35 per ton to get the ores there on pack animals.

In 1880 our ore began to go to the Pueblo works of Matter & Gets, by wagon to Alamosa, the distance to the latter place being over 200 miles, and the cost of freight and treatment of our ore to the Pueblo Smelting and Refining Company is now $35 per ton. In addition to this the cost of supplies is reduced at least 50 per cent, and the cost of packing the ore from the mines to Ouray at least $5 per ton, so that the average cost of producing the ore and getting it to the shipping point is at least $10 per ton less than in 1880. Therefore we save $20 per ton in freight and treatment to Pueblo, and $10 per ton in producing and getting the ore to Ouray, which makes us receive $30 per ton more for our ore now than we did in 1880. For all this we are indebted to the Denver & Rio Grande Railway, and as the same figures will apply more or less to all the camps on the San Juan region, the incalculable benefit conferred on the State of Colorado by mines of a thousand prospects whose ores three years ago were as so much waste. If the Denver and Rio Grande Railway built a branch from their main line at Montrose to Ouray, the further reduction on freight of ore will not be very great — perhaps $5 per ton — but we shall be able to ship all the ore we can produce, whereas now we are dependent on such wagon freight as may be available. In addition to this, the reduction in prices of supplies and labor and the building of wagon-roads to the mines will reduce the cost of laying the ore down in Ouray at least $10 per ton more, which will give us $45 per ton more for our ore than we received in 1880; and it must be remembered, too, that this is reckoning in the ore going to Pueblo, a distance of 275 miles. It does seem to me that with extensive coal beds in the Uncompahgre Valley, and the mining camps of the Dolores, San Miguel, Ophir, Sneffels, Uncompahgre, Red Mountain, and Poughkeepsie tributary to it, there should be a smelting center established in that valley some time in the near future.

To give an idea of the decrease in prices, in 1877 I paid $1 per pound of giant powder that I now buy for 35 cents, and $12 per 100 pounds for flour that I now buy for $4.50. Hay has come down from $55 per ton to $25 per ton, and oats and corn from 10 cents per pound to 3 cents, potatoes from 5 cents per pound to 1-½ cents, and other things in proportion. And yet in the face of such outrageous prices as we paid for everything in the early days, and the fact that we were 200 miles from a railroad, there were always men to be found outside who would meet you with the following idiotic question: "If your mines are as rich as you say, how is it that you haven't produced more bullion?"

Another assurance of our permanent prosperity has been furnished in the fact that concentration, or getting rid of the waste rock in our ores by machinery, has been proved a success. There are thousands of tons of ore lying on our dumps which will go $50 to the ton and which up to the last year has been considered so much waste, but with machinery that will reduce this, say, six tons into one and save say even 80 per cent of the assay value, we have instead of six tons of waste one ton of $240 ore. And still another point that I have omitted is this — we now get paid for the lead and copper in our ores and also for the fractional parts of an ounce of gold per ton, all of which we got nothing for two years ago.

Since commencing this letter, I read in the *Dolores News* as follows: "For two days we have had no eastern mail on account of the snow-bound train on the Conejos Range," and in another part of the paper a long editorial advocating a northern railway outlet for Rico via Trout Lake, Ophir, Ames, and San Miguel to a junction with the proposed Ouray branch of the Denver & Rio Grande Railway at the forks of the Dallas, and thence on to the main line at Montrose; and the following are the very strong and sensible reasons given, in the same paper of January 20:

"One of the chief advantages of such a line is that it would place Rico in direct communication with the western markets, Salt Lake and San Francisco, as well as being the nearest route to Pueblo and Denver and the East, as the Denver & Rio Grande will soon be running through cars from Denver to Salt Lake via Montrose. Again the Gunnison Valley with its inexhaustible supplies of coal, fuel, and fluxes, surrounded as it is, with rich mining camps from Rico and Gothic, and from Tin Cup to Sierra La Sal, must inevitably become the great ore reducing and commercial center of the State west of the Continental Divide. Already this great valley has two competing railroads, and there it is that Rico must look for competition in the hauling as well as the treatment of its ores. Again, this line would place Rico in immediate connection with its neighboring mining towns, a point not to be overlooked: for it is to the mining towns that Rico must look for co-operation and mutual assistance in all matters of general progress and public improvement, and not to its neighboring ranch towns which possess the sole virtue of being the center of everywhere." *(In 1892 the Rio Grande Southern Railroad was completed from Ridgway to Durango, giving Dolores, Rico, and Telluride access by rail to the outside world. However silver prices dropped drastically the next year, which closed many of the mines that had been waiting for cheap transportation.)*

So that with all the above camps united in presenting their claims for a branch of the Denver & Rio Grande main line, we have a very good chance of getting it at once.

The tax list shows an increase of 50 per cent in the assessed valuation of the county of Ouray this last year over the pervious one, and this present year will, I think, more than sustain a similar increase.

SHERIDAN

(From *The Denver Republican*, February 19, 1883)

The Sheridan, in the San Miguel District, is a full claim, 1,500 feet in length by 300 feet in width; title, United States patent. Up to the spring of 1882 little more than assessment work was done on this lode. Last summer, however, $25,000 worth of ore was produced from the following workings: a drift or tunnel was run on the vein 100 feet in a northerly direction, and 10 fathoms above it another drift 25 feet in length and a winze was raised from the lower drift to the end of the upper one. A slope, about 80 feet in length and 36 feet in height, was then taken out between the two drifts. In addition to the $25,000 worth of ore sold, there remains on the dump, sacked and ready for packing to the concentrator, 250 tons of low-grade ore. The average width of the ore-body in the above workings is 2 feet. At this writing (February 1), 14 men are being employed at the mine, continuing both levels aforesaid, and taking out six stopes — 36 feet in each. The product of this latter work during these winter months (say 'till May) will be 150 tons of $200 ore, 100 tons $300, 50 tons of 500 ounce ore, and 300 tons of concentrating ore that will average about $60 per ton.

The ore has steadily increased in its yield of silver and gold as depth was gained by the workings, and this week the workmen have encountered a chimney of ore, which surpasses anything yet seen in that phenomenally rich vein. I have it from a gentleman just from the mine and whose statements are perfectly reliable, that this new ore-body is four feet in width and will mill from 300 to 500 ounces to the ton. Of course, if this chimney holds any distance the winter's output will be materially increased over the figures already being given.

The Sheridan is now being worked under lease by Mr. D.C. Hartwell, of Ouray.

THE SIDNEY

The Sidney, in the Sneffels District, is an extension of the Terrible, and is owned by the Emma Mountain Mining Company, David Geiger, of Ouray, manager. The work done consists of a cross-cut tunnel, 546 feet in length, which intersected the vein at a depth of 250 feet below the surface, encountering a body of very fine argentiferous galena, varying in width from 1 to 2 feet. This ore yields from 50 to 150 ounces of silver to the ton and 35 to 50 per cent lead. They are now drifting both ways on the vein and will stope ore in the spring.

BELL BROTHERS STRIKE

On the first of February I examined what is known in the camp as the "Bell Bro.'s Strike," although a great many

THESE MINERS POSED AT 12,800 FEET ON THE DUMP OF THE SIDNEY MINE IN SIDNEY BASIN IN THE SNEFFLES DISTRICT. THESE ARE NOT GENERALLY THE MEN WHO PROSPECTED, BUT RATHER THE HARDWORKING MEN WHO HELPED DEVELOP THE MINES.

Courtesy Western History Dept., Denver Public Library. X-61113

others have staked claims in what appears to me to be a continuation of the same deposit. We are all of us too prone to judge by appearances of rock to what its yield of the precious metals is, or ought to be, and every day we hear of neglected rock turning out rich. *(Isn't it ironic that this happened to Weston himself with the Camp Bird Mine!)* The existence of mineralized quartz and quartz-porphyry where this late discovery has been made has been known since '76, but no one seemed to think it worth testing or exploring; for it is only lately that flat deposits have been deemed worth anything, and in the "early days" here, we would stake nothing but vertical fissures that outcrop everywhere, and none but those that "ran high" on the surface. Now, this ore has always been deemed to be everywhere in general and nowhere in particular, and so it was not worked until Bell Brothers, who are practical miners and assayers, made some tests of apparently worthless rock, with astonishing results.

Another fact that is turning our miners' attention to large deposits of ore of an average low grade is the fact that our ores are being concentrated successfully; and the late strike will furnish vast quantities of ore, which appear to me essentially adapted for concentration.

It is situated in what is known as the Amphitheater, on the east side of town, the altitude being 500 feet above the main street of Ouray, and only about fifteen minutes walk from it. As it was in December, and a good deal of snow on the ground before it was discovered to be good ore, no work was done of any importance, but as far as I can judge from what I saw, I think it is a flat deposit on a lime-stone bed, very similar to the Mineral Farm. The ore appears to be a mixture of copper pyrites, galena, iron pyrites, blue and green carbonates of copper, and oxide of manganese, associated with calcspar, the principal gangue or matrix being quartz porphyry, and the barren rock amongst it an altered sand-stone. Above this deposit has at one time been volcanic rock, I think, which has been eroded or worn off, and it is therefore a genuine contact deposit between sedimentary and volcanic rocks. At the foot of the hill are numerous hot springs, and it seems rational to suppose that these, having the minerals in solution, have passed over the limestone bed, the limestone having the effect of precipitating the minerals, and the subsequent volcanic overflow did the rest by heat and pressure.

The thickness of the ore-bed is apparently about nine feet. The rich ore is contained in the rock in small bunches or pockets distributed unevenly through the rock. It cannot be hand-sorted, but will have to be cobbed with sledge-hammers and then put through the concentrating mill.

The following assays of ore were made by the Bell Brothers:

Silver, 456 ounces; gold, $25
Silver, 148 ounces; gold, $20
Silver, 175 ounces; gold, $35
Silver, 85 ounces; gold, $5
Silver, 161 ounces; gold, $5
Silver, 159 ounces; gold, $5
Silver, 196 ounces; gold, $20
Silver, 397 ounces.

Mill runs of average samples:

Silver, 198 ounces.
Silver, 110 ounces; gold, $6

The first assay the Bell Brothers made was from a big boulder, 10 pounds of which, broken off, yielded at the rate of 248 ounces silver to the ton.

Colonel Todd has a claim in the same deposit, called the Monday, an assay from which yielded at the rate of 382 ounces silver and $20 in gold to the ton.

What the actual average yield of this apparently vast deposit is, we shall not know till the Bell Brothers have run a few tons through their concentrator, which will be finished this summer; but Mr. Bell informed me that the very lowest average they have been able to get is 20 ounces silver per ton. Now, if the general average should only be this much, it is good enough, for it can be mined and concentrated for $10 per ton, at the outside, and should concentrated, in my opinion, at least 10 tons into 1. But with the tests made so far, it would seem that the general average will be a good deal higher than 20 ounces. I should have mentioned that, although the principal nature of the ore is quartz porphyry, there is a good deal of genuine quartz mixed through it, with pockets of drusy quartz and beautiful large crystals.

Bell brothers own four claims, the Silver Crystal and Quartzite No. 1, 2, and 3; Dr. Rowan, of Ouray, the Emma W. and Hannah; Colonel Todd, the Monday, and there are a great many more whose names I could not ascertain.

COAL

About twenty-five miles from Ouray, and within six or seven miles of the Denver & Rio Grande Railway, is a coal field almost entirely unprospected and undeveloped, but, from the surface outcrop, it is known to cover an area of at least 2,000 acres. Where it outcrops on the side of the mountain there are five strata of bituminous coal, having a respective width of 4, 6, 20, 8, and 9 feet; so far as tested it is of excellent quality, and makes but a small percentage of ash. This only partially explored coalfield is on the divide, between the Cimmarron and the

Uncompahgre, and its enormous importance to Ouray County seems yet hardly to be realized. *(This large coalfield in Cow Creek has never been developed.)*

About one hundred miles from Ouray are the great coalfields of Gunnison County, which are said to cover an area of 15,000 square miles, 300 square miles of this, at the headwaters of the north and south fork of the Gunnison, being anthracite. *(These coalfields were extensively developed and were mined for almost seventy-five years.)* The following is from the *Elk Mountain Pilot:*

"The coal interest, which is still in its infancy, so to speak, is bound to be a great resource of wealth to this country. We all know that coal or iron mines are comparatively valueless without the aid of a railroad, and wherever the railroad has approached the coal fields it has been turned into a beehive of busy workers. The Rio Grande has reached a point four miles above Crested Butte, where it is terminated right into a large coal field, and where works are being erected to crush the coal for market. At Crested Butte the Colorado Coal and Iron Co. have been operating for over a year shipping coal and burning coke. To one who is acquainted, the approach to Crested Butte reminds you very much of Pittsburg, with the heavens lighted up from the burning coke beds. From Crested Butte to Irwin, where the anthracite coalfields begin and run west, is a distance of eight miles. The route has been surveyed and partially graded, and we may look for the road to be built here most any time.

From the Denver papers we find the following figures for the year's shipment.

Soft Coal — C.C. & I. Co., Crested Butte, 40,000 tons; Skinner & Jefferson, Crested Butte, 1,000 tons. At

Mount Carbon, the Richardson's bank has shipped 1,500 tons; South Park, 1,000 tons, and small shipments from other banks amounting to 500 tons, making a total of 44,000 tons for the year.

Anthracite Coal — From Crested Butte, the Messa Company have shipped 1,000 tons, M. Rush Warner, 500 tons. At Irwin, the Thompson Bank has shipped about 800 tons, making a total of 2,300 tons. Of course, that from the Thompson Bank has only been mined for domestic use in Irwin, there being no railroad to it, not a pound had been shipped to any other market.

Here is an industry that requires large capital and labor to carry it on. Two years ago the coal industry was hardly known and there were no railroads in the country, but today there are two railroad corporations steaming up and down the rugged mountain defiles, while figures show that there were 26,300 tons of coal mined and shipped. Can anybody tell what will be the increase in another two years? The magnitude of the coal industry of Gunnison county is beyond conception, and we can only await the progress.

COLORADO

HER OUTPUT FOR 1882.

The following is clipped from the *Denver Republican:*

The grand increase of the year has occurred in Lake County and the San Juan country, the former, owing to the better development given properties, and the latter to the facilities of railway connection and the ability to ship ores by rail that would not bear wagon transportation. Parts of Gunnison and Summit counties will emulate the record made by San Juan, and notably the camps of Gothic and Irwin, in Gunnison; Chihuahua, Decantur, Montezuma, and others, in Summit.

The shrinkage in a few of the counties is to be credited to pending sales, the testing of mills for treatment of the product on the ground, or development work, which prevents outputting of ore. Of this class, the Great London Mine, of Park County, is a shining example of immense ore bodies held in reserve until the owners are ready to treat their product.

Want of space prevents the *Republican*, in this issue, from a further review and analysis of the many features of interest to which attention might profitably be called in the subjoined tables, and the reader must enter into these calculations for himself.

The counties that may consider their goals too small can, after a more searching investigation than can now be made, avail themselves of the columns of the *Republican* to make such corrections as the facts shall justify.

Reference to the figures massed below will give any one of an investigating turn of mind all the special employment he may desire for some time to come.

MINING PRODUCT OF COLORADO FOR 1882.

County.	Amount.	County.	Amount.
Boulder	$ 550,000.00	Lake	$17,131,853.00
Chaffee	225,500.00	La Plata and	
Custer	705,116.00	San Juan	675,000.00
Clear Creek	2,001,629.00	Ouray	329,760.00
Dolores	125,000.00	Park	283,564.00
Fremont	19,960.00	Pitkin	100,000.00
Gilpin	2,006,516.00	Rio Grande	310,000.00

Grand	10,000.00	Routt	100,000.00
Gunnison	600,000.00	Saguache	52,000.00
Hinsdale	275,000.00	Summit	1,150,000.00
		Total	$26,750,898.00

MINING PRODUCT OF COLORADO FOR 1881.

County.	Amount.	County.	Amount.
Boulder	$ 535,482.88	Lake	$12,738,902.00
Chaffee	196,400.00	La Plata and	
Custer	755,600.00	San Juan	40,000.00
Clear Creek	2,204,980.34	Ouray	78,000.00
Dolores	125,000.00	Park	350,000.00
Fremont	14,535.50	Pitkin	120,000.00
Gilpin	2,150,700.00	Rio Grande	280,000.00
Grand	10,000.00	Saguache	53,500.00
Gunnison	535,033.00	Summit	1,828,000.00
Hindsdale	187,375.00		
		Total	$22,203,508.72

The *New York Mining Record,* in an editorial headed, "Colorado as a Factor in the Field of Mining Industry," after commenting on the increased mineral product of the State, says as follows:

"It is also claimed that the value of the coal output of the State at two dollars per ton at the mines has reached the sum of $4,460,000, including 10,000 tons of anthracite from Gunnison County. The coke production for the year is estimated at $450,000, with every promise of a materially enlarged product for 1883, with a ready market in New Mexico and Arizona as well as at home."

In iron mining and the fabrication of iron and steel, the people of Colorado have another source of congratulation, for during the early part of 1882 Bessemer works were started, and nearly 17,000 tons of steel rails have been turned out, with many descriptions of manufactured iron and steel, bar for rails in mines, 22,000 kegs of nails, and 21,000 tons of steel ingots.

Coupled directly and consequently with these encouraging results of the mining industry of the State are the agricultural and pastoral achievements for the year. In actuality, these are well nigh as much the consequences of the money and labor employed in Colorado in the development of the subterranean riches of the State as the gold and silver, copper and lead to the value of $25,000,000 of more which the hard-handed labor of the miners has extracted during the past year.

It is because of the mineral resources of Colorado and their steady development during the past five years that there are today 3,088 miles of railway in the State, of which 880 miles have been constructed during the last year, and of which nearly 2,000 miles have been built since the remonetization of silver in 1878. From the same cause, almost as directly has it come to be that there are about one million sheep in the State, worth $2,500,000, giving a wool clip last year worth $1,000,000, while furnishing meat worth $350,000. From the same influence the wheat crop of the year reached 1,533,740 bushels, oats 1,083,287 bushels, corn 600,00 bushels, and barley 261,185, and that the plains and valleys of the State swarm with herds of cattle assessed at $5,435,616, and that as much as $3,138,000 is the assessed value of the horse property in the State. From the past year's aggregate of mining results in Colorado should be drawn a pregnant lesson as to the vast importance to the country of its possession of so much mineral wealth awaiting development.

ADENDUM.

Among the valuable properties in the Uncompahgre District is the Chicago Mine, an extension of the Grand View, owned by Costigan, Terpening & Buckley.

The Silver Star, on the Silver Link vein, owned by Costigan, Crummey & Story and also:

The Sneffles No. 2 lode, Upper San Miguel District, Marshall Basin, owned by Leeson & Costigan.

The Belmont Lode, Marshall Basin, upper San Miguel District, owned by Costigan & O'Neal.

Since the MS. For this pamphlet was sent to the printer, the legislature has created a new county of San Miguel, and some of the mines described in this book as Ouray county mines are now in the county of San Miguel.

OURAY BUSINESS DIRECTORY.

Grocers.

A. W. Begole.

M. L. Corbett & Co.

Chas. H. Rawles.

Buenemann Bros.

Hardware.

Alling & Co.

Gordan Kimball

Dry Goods and Clothing.

Dutton & Siddons.

Men's Furnishing Goods and Clothing.

Howe G. Corson.

Bakery, Fruit and Oysters.

Leo Erdmann.

Bakery and Restaurant.

H. Copp.

Boot and Shoe Store.

E. C. Meyer

Chemists and Druggists.

Scott & Ashley

Hazard & Rowan

Furniture, Glass Ware, etc.

Haskins & Kilpatrick.

Assayers.

Munn Bros.

F. C. Meyer.

Bell Bros.

Ouray Sampling Works.

Munn Bros.

Insurance Agent.

W. W. Rowan M.D.

Civil and Mining Engineers and Deputy U. S. Mineral Surveyors.

Reed Bros.

Chas. A. Wheeler.

Agent, Pueblo Smelting and Refining Company and Forwarding and Commission Merchant.

D. C. Hartwell

Saw and Planing Mills.

Orton & Mackoy.

Munn Bros.

Photographer.

J. E. Gilbert.

Millinery Stores.

Miss Wemple.

Mrs. Waring.

Wholesale Butchers.

Strayer & Mark.

Retail Butchers.

Parliman & Gray.

J. G. Boedecker.

Saloons and Billiard Rooms.

Wiggins & Patrick.

Nevill & Nankervis.

Pat. Hess.

W.E. Fay.

George Palmer.

Mayold & DeFrance

Yost & Englehart.

Brewery.

Ouray Brewing Company.

Hotels.

Dixon House, A. DeFrance

Grand Central, A. G. Williams

Restaurants.

Mrs. Marsh.

H. Copp.

Abstract Office.

C. W. Haskins.

Blacksmiths and Coal Dealers.

W. L. Finch

Louis King

Livery Stables.

Louis King & Co.

Nutter & Clark.

Saddler and Harness-maker.

John Long.

Laundresses.

Mrs. G. H. Moore.

Fanny Battle

Chinese Laundry.

Sam Pang Lee Fong.

Boot and Shoe-maker.

Geo. Allen

Carpenters.

Jas. Coumbe.

Shank & Waring.

Hot Mineral Spring Baths.

Mrs. W. Buchanan.

Barber and Hairdressers.

John Carney.
W. Fulkerson

Lawyers.

Wm. Story
Hon, Jerold R. Letcher.
Costigan & Stewart
Cassidy & Cobb.

Surgeons and Physicians.

W. Ashley, M. D.
Alex. Hazard, M. D.
W.W. Rowen, M. D.

County Officers.

COUNTY COMMISSIONERS.

C. Hartwell, Chairman
John Merling.
Milton Evans.
Judge . . .Theron Stevens.
Sheriff . .Adelbert Parsell.
Treasurer . . . . .W. W. Stoddard.
Clerk and Recorder . . . . . .Chas. Haskins.
Assessor . . . . .A. E. Long.
Clerk Dist't Court . . . . .J. Stuart Meyers.
Coroner . . .W. W. Ashley M. D.
Sup't Public Instruction . .W. W. Rowen M. D.

Town Officers.

Mayor . . .W. W. Ashley M. D.

TRUSTEES.

Gordan Kimball.
Theo. Hess.
John Worthington.
A. C. Yost.
Town Clerk . . . J. Stewart Meyers.
Police Magistrate . . . . .A. L. Sivyer.
Chief of Fire Dep't. . . . . . .Howe Corson.
Chief of Police . . . . . .Luther Harris.

POSTMASTER.

A. G. Dunbar.

Bank.

Miners' and Merchants' (Thatcher Bros.) J. M. Jardine, Cashier.

Forwarding and Commission Merchant, Fast Freight Line between Montrose to Ouray.

David Wood

Real Estate and Mining Broker.

G. Patterson.

W. Weston, miner in San Juan for six years, four years' underground experience, holds assayers' certificate from the Royal School of Mines, London.

Will report on mining properties.

Glossary of Some Mining Terms

(Adapted from Frank Fosset's *Colorado*)

Adit — A tunnel on a vein or lode; a passage for water underground.

Amalgam — Quicksilver mixed with other metals, such as gold and/or silver.

Arastra — A Spanish ore reduction machine consisting of a round hard platform, upon which the ore is deposited and crushed by means of a revolving sweep to which huge flat stones are attached.

Bed rock — The bed of a metalliferous deposit.

Blind lode — One that shows no surface croppings.

Blossom rock — Float ore, found upon the surface or near lodes or ledges outcrop, and from which they have become detached.

Breast — The face or head of a drift.

Cage — The elevator used for hoisting and lowering the ore, men, or materials in a mine.

Country Rock — Rock on either side of a lode or ledge, usually barren, which encloses a vein.

Cap Rock — The formation of rock overlaying the "pay-dirt" ore.

Carboniferous — Containing coal.

Chute — An incline or opening from one level to another, through which to slide or pass ore.

Conglomerate — "Pudding stone" composed of gravel and pebbles cemented together.

Contact — A lode lying between two different kinds of rock.

Croppings — Ledge matters lying on the surface or the outcropping of a vein.

Crosscut — A drift run at right angles to the ledge, for the purpose of ascertaining its width and to otherwise prospect it; also an opening or level driven from one vein to another.

Cupriferous — Containing copper.

Debris — Sediments from hydraulic or other mines and not containing ore, usually piled below the mine's opening.

Detritus — A mass of substances worn from solid bodies by attrition and reduced to small particles.

Development — Work done to open a mine and reach the ore.

Dead Work — Work done to get to or discover ore but which produces no ore itself.

Drift — Tunnels leading off from the main shaft or tunnel following through or along a vein.

Dump — The place where ore or waste rock is deposited after being taken from the mine.

Face — The extreme end of a tunnel, drift, or excavation.

Float — Fragments of a lode that have been moved by water or other means and left on nearby surfaces. A vein or ledge can often be found by following float back to its source.

Flume — Boxing or pipe for carrying water.

Flux — Any substance employed to promote the fusion of minerals.

Foot Wall — The lower wall of a lode or vein.

Free Gold — Gold easily separated from rock or dirt.

Gangue — The matrix; quartz is commonly called the gangue rock of gold or silver ores, though it might be other materials.

Grizzly — Bars set in a flume to screen out the large rocks.

Hanging Wall — The upper wall on the lode or vein.

Horse — A mass of wall rock or other barren matters obtruding into a lode or fissure.

Incline — A slanting shaft.

Ledge — Synonymous with lode or vein.

Level — Drifts from the main shaft or from one shaft to another; usually at set distances of every 50 or 100 feet.

Lode — A longitudinal fissure or chasm filled with ore-bearing matter and giving well-defined side walls; a mineral vein in country rock. Vein, lode, and ledge are synonymous.

Mine — Territory on a lode or vein which generally consists of shafts, levels, tunnels, or adits, winzes, and stopes, together with machinery and surface buildings. There may be several mines on one vein or lode.

Placer — Alluvial deposits; earth containing gold deposits.

Plant — The interior workings of a mill.

Porphery — A barren ore, stratified and in which are imbedded crystals.

Primary Rock — Various kinds of slate, granite, quartz of irregular crystalline matter.

Prospecting — Hunting for lodes or veins.

Panning — To wash dirt from free gold with a pan.

Pulp — Pulverized ore; generally very fine.

Reduction — Separating from foreign substances; extracting metals from the ore.

Reserves — Ore reserves are the vein material still standing in the mine.

Salting — The act of introducing rich ore into a mine for deceptive purposes.

Schist — Clay slate.

Secondary Rock — Those above primary and below tertiary; usually sand, clay, and pebbles cemented together.

Shaft — A vertical or incline excavated for prospecting or working a mine.

Strata — A flat bed or layer of rock, earth, or mineral.

Stope — Breaking the ore overhead or underneath or on the side, creating a "room" empty of ore.

Stoping — Breaking ore from a stope or section of ground in a mine.

Stull — Platform of timber between levels for strengthening the mine by supporting the walls.

Strike — A find of valuable mineral.

Sulphurets — Combinations of sulphur with a metallic, earthy, or alkaline base.

Tailings — The auriferous earth that has been washed and deprived of the greater part of valuable minerals; a sandy type material left over after ore is run through the mills.

Tertiary Rock — Rock lying above the secondary and below drift, usually sandstone, clay limestone, sand, etc.

Trend — The course of the vein.

Upraise — Running a drift upwards.

Winze — A shaft connecting on drift or level with another, but not reaching to the surface (which would make it a shaft).

Wall — Boundary of a vein.

INDEX